A-Z
INORGANIC CHEMISTRY

A-Z INORGANIC CHEMISTRY

Prof. Nausad Ali

CENTRUM PRESS
NEW DELHI-110002 (INDIA)

CENTRUM PRESS
H.O.: 4360/4, Ansari Road, Daryaganj,
New Delhi-110 002 (India)
Ph.: 23278000, 23261597
B.O.: No. 1015, Ist Main Road, BSK IIIrd Stage
IIIrd Phase, IIIrd Block,
Bangalore - 560 085 (India)
Tel.: 080-41723429
Visit us at: www.centrumpress.com

A-Z Inorganic Chemistry

First Edition, 2009
ISBN 978-93-80106-98-4

PRINTED IN INDIA

Printed at Mehra Offset Press, Delhi

Contents

Preface

Inorganic chemistry is the branch of chemistry concerned with the properties and behavior of inorganic compounds.

The Basics of Inorganic Chemistry is an effort to provide an account of Inorganic Chemistry not only suitable as an introduction to this field but also provide a sequential compendium of all the latest developments in the field, whereas the selection of subject matter for the present set has been chiefly according to the syllabi prescribed by the Universities for the students of Chemistry.

This book should therefore be useful in an advanced undergraduate course in Chemistry.

Author

Chapter 1

Hydrogen

Hydrogen has been receiving interest as a “source of energy” recently. Besides, it is interesting in itself, so its nature and properties will be discussed here. Hydrogen is such a simple system that it has been important in the development of physics, a test bed for quantum mechanics, which it has amply confirmed. When we talk about hydrogen, we must distinguish between the hydrogen atom, hydrogen gas (the diatomic molecule), the proton (its nucleus), and the hydrogen that is part of many chemical compounds.

Hydrogen is apparently the most common substance in the universe, because it has no competitors for the title, except perhaps for helium. It exists mainly in the interstellar medium, with an average density of 10 atoms or less per cubic centimetre, and is by no means uniformly distributed. It has no effect on visible light, and was only discovered when ultraviolet observations could be made outside the earth’s atmosphere, since it absorbs the far-ultraviolet Lyman alpha line at 12.15 nm. It also emits radiation of wavelength 21 cm, from a transition between hyperfine levels of its ground state, that has also been observed. Hydrogen molecules do not seem to have been observed in the interstellar medium, but hydrogen atoms are generally taken as indicating the existence of a wide variety of other molecules.

Hydrogen exists on the Sun, because its absorption lines are observed in the solar spectrum. Although there are hydrogen atoms in the solar atmosphere, hydrogen itself does not exist in the Sun, but only a plasma of protons and a few other nuclei, rendered neutral by a sea of electrons. If these protons are called hydrogen, then hydrogen is by far the most abundant element in the Sun, with helium second and everything else much rarer. The Sun's energy is produced by fusion of protons to form helium (alpha particles), in the proton-proton chain. Two protons react to form deuteron, consisting of a proton and a neutron, after emitting a neutrino and a positron. A proton reacts with a deuteron to form helium-3, which has two protons and one neutron, emitting a characteristic gamma-ray. Two helium-3 nuclei then react to form an alpha particle, or helium-4 nucleus, and two protons. A temperature of 10^7K is necessary to achieve the kinetic energies to bring the positively-charged particles close enough together to react. 90% of the Sun's energy comes from the proton-proton chain, and 10% from other fusion processes. The neutrinos released should be observable on earth, but at first seemed to be absent. Recently, it was discovered that they changed from electron neutrinos to muon neutrinos on the way to the earth, and then would not be detected by counters of electron neutrinos. The absence of the characteristic gamma rays in the cold-fusion claims some years ago was a sure sign that no fusion was going on. These gammas are absorbed deep in the Sun, of course, fortunately for us.

The Sun emits a plasma consisting mainly of protons and electrons at high speed in all directions, which is called the *solar wind* and can be considered as the outer part of the Sun's corona, rushing outward at 400 km/s or so from areas of weak magnetic fields, called corona holes. This plasma interacts in complicated ways with the magnetic field of the earth to give the aurora borealis, magnetic storms, and other effects. Comets are surrounded by a cloud of hydrogen that can be as large as the Sun. This cloud results from the decomposition of water by ultraviolet radiation.

On the Earth, hydrogen is not very common, existing only in the thin sheet of water covering 70% of the Earth, as only 0.127% by weight of the lithosphere, and in the small amount of hydrocarbons and organic matter. It is present only to about 5×10^{-5}% by volume of air (about the same as nitrous oxide, and less than methane or carbon monoxide). It tends to become more important at very high altitudes, rising to about 1% of a very, very rare atmosphere. The hydrogen atoms at high altitudes may be the result of the solar wind, as well as the hydrogen that has diffused from below, and all this hydrogen is gradually lost, since the Earth's gravity is not sufficient to retain it. Hydrogen gas is present in small amounts in volcanic gases where it is produced by chemical reactions, but more of it appears combined as hydrogen sulphide and nascent water. There are no practical sources of hydrogen gas. All that we use must be manufactured with the expense of energy from hydrogen compounds.

The hydrogen atom is composed of a very light, negatively charged *electron*, and a heavy, positively charged *proton*. The weight of the proton is 1.00758 atomic mass units (about 6.02×10^{23} protons weigh 1.00758 grams). The electron weighs about 1/1837 of the proton's weight. The diameter of a proton has a fairly definite meaning, and is 1.6×10^{-13} cm. The diameter of a hydrogen atom in its ground state is around 1×10^{-8} cm, so the ratio of the diameters is about 1.6×10^{-5}, which makes the ratio of the volumes 4.1×10^{-15}. The volume occupied by the electron is all but a tiny fraction of the volume of the atom, while 99.95% of the atom's mass is in the proton.

The question of the "discovery of hydrogen" only has meaning after the recognition of the significance of chemical elements in the 18th century. As soon as strong mineral acids were discovered, dissolving metals in them often produced bubbles of hydrogen. Paracelsus could very well have made some hydrogen, and even made it explode, but there was no recognition of the significance of these observations, or of the real nature of gases. Henry Cavendish first studied the gas in 1760 or 1766, and showed that it burned to water with Priestley's oxygen. Lavoisier named it hydrogen, "water-former," in 1783, the first year that it was used in balloon ascents.

THE HYDROGEN ATOM

The hydrogen atom may be considered as a mobile package of an electron and a proton, available for any adventures it may encounter. Two atoms can easily come together, since they do not repel one another, and when they do, they find that it is much more comfortable for the two protons to move close together (to 0.7416 Å) and arrange the electrons around themselves to achieve the minimum energy. This forms a hydrogen molecule, H_2, whose dissociation energy is 4.476 eV, the energy of a 277 nm far-ultraviolet photon. Since there is radiation of this and smaller wavelengths in space, and collisions are rare, we can expect all the hydrogen there to be atomic. On the Earth's surface, however, protected from ultraviolet and at higher pressures, the diatomic molecule is stable.

There is a paradox here. Although atomic hydrogen is very active, diatomic hydrogen is not; it is quite inert until the molecule is disrupted. There is so little attraction between molecules that hydrogen melts at -259.20°C (14K) and boils at -252.77°C (20.4K), lower temperatures than for any other substance except helium, which we may expect to be inert. The critical temperature is -240°C (33K), the critical pressure 12.8 atm, and the critical density 0.0308 g/cc. Hydrogen is very difficult to liquefy, but Dewar managed it in 1898. In spite of this, hydrogen can explode when mixed with oxygen, the very opposite of inertness. The secret, of course, is *reaction mechanism*.

If just one hydrogen atom H* (an excited state is represented by the *) should be present, then the reactions

$H^* + O_2 \rightarrow HO^* + O^*$; $O^* + H_2 \rightarrow HO^* + H^*$; $HO^* + H_2 \rightarrow H_2O + H^*$ produce additional H* atoms, which then go on to react further. This is a *chain reaction* that proceeds more and more rapidly, gobbling up the H_2 and O_2, both of which are otherwise inert. Hydrogen is inflammable over the very wide range of concentration from 6.2% to 71.4% in air. The ignition temperature is also low, only 580°C to 590°C, 70°C lower than for methane. If a little air leaks into a volume of methane (natural gas) there is little hazard, since mixtures over 13.3% are not inflammable. The same is true of gasoline. If a little air leaks into a volume of hydrogen, the hazard is great, for then the gas will explode, not just burn. Hydrogen is not very soluble in water; at 0°C, 100 ml of water dissolves 1.93 ml of hydrogen.

In energy, according to $E = mc^2$, 1 amu is 931.16 MeV, so the energy equivalent of the proton's mass is 938 MeV. The energy equivalent of the electron's mass is 1/1837 of this, or 0.511 MeV. The ionization energy of a hydrogen atom, the energy required to free the electron from its ground state, is 13.5 eV. The electromagnetic energies released or absorbed as the electron moves between its stationary states in the field of the proton comprise its *spectrum.* The hydrogen spectrum is by far one of the simplest, but its explanation defied classical physics. In the visible and ultraviolet, only a series of lines, called the Balmer series, were observed. The longest wavelength is 656.3 nm, called Hα, and then 486.1, 434.0, 410.2, 397.0, 388.9, and so on, approaching a limit at 364.6 nm. These could not be explained as harmonics of any fundamental frequency, nor were harmonics of these frequencies observed, as would be the case in any mechanical system. The series is named after Balmer because he found a simple empirical formula, $\lambda = Cn^2/(n^2 - 4)$ that gave the wavelengths of every member of the series.

In 1911, however, Bohr created a simple mechanical model of the hydrogen atom, but with the restriction that angular momenta were quantized in units of $h/2\pi = h'$, where h was Planck's constant from the theory of radiation. This gave a series of *stationary states*, and it was postulated that the frequencies of the spectral lines were proportional to the differences in the energies of these stationary states, where $\Delta E = hf$. Since the energies of the stationary states were given by $E(n) = -R/n^2$, where n = 1,2,3,..., $f = R(1/n^2 - 1/n'^2)$. For $n < n'$, there is emission of radiation, and for $n > n'$, absorption. The Balmer series in emission is given by n = 2, n' = 3, 4, 5,..., or by a "jump" from higher states to the state n = 2. This formula can be modifed to allow for a nucleus of any charge or mass.

The ground state is n = 1. The far ultraviolet series corresponding to n' = 2, 3,... is the Lyman series, of which the longest wavelength is 121.5 nm. The least excitation of a hydrogen atom in the ground state is from n = 1 to n = 2, which corresponds to this line. Under normal conditions, hydrogen is in its ground state, and so does not absorb radiation of longer wavelengths. Also, far ultraviolet is strongly absorbed by the atmosphere.

The lines of the Balmer series are seen in absorption in the solar spectrum. Hα is the C line, Hβ the F line, and Hγ the G line. The photosphere emits a continuous spectrum that is modified by the relatively cooler atomic hydrogen gas in the chromosphere. The appearance of the Balmer series means that many atoms are in the first excited state, n = 2, in the chromosphere. The Balmer series is strongest in stars of spectral class A, weakening in hotter or cooler stars. The Sun is class G, where lines of ionized calcium are prominent, but the Balmer series is still seen.

Bohr's theory has no firm foundation; it is merely a remarkable coincidence. Quantum mechanics put it on a firm foundation after 1925, however, and verified all its predictions. One fault of Bohr's theory was that it could not be extended to the case of more than one electron, or to molecules. Quantum mechanics gave excellent, though approximate, results for helium, with two electrons around a nucleus of charge Z = 2, and for molecular hydrogen, H_2.

A proton and a neutron combine to form the stable *deuteron*. Adding an electron makes heavy hydrogen, or *deuterium* D, with an atomic weight of 2.014. D_2 has mp -254.60°C, bp -249.7, a little higher than H_2. Deuterium was discovered by H. C. Urey in 1931 or 1932. It makes up about 1 part in 5000 of natural hydrogen, and its chemical behaviour is identical. Two neutrons can also combine with a proton to form a *triton*. The triton is not stable, but emits an electron of 18 keV peak energy with a half-life of 12.26 years to become He^3. The corresponding atom is called *tritium*. Tritium is produced only by nuclear reactions, including those of cosmic rays, and its abundance is no greater than 1 ppm.

The free neutron, incidentally, is also radioactive with a half-life of 13 minutes, emitting an electron of peak energy 0.78 MeV to become a proton. That is, it decays into hydrogen.

Neutrons in a nucleus do not do this, since they do not get bored there.

THE 21-CM LINE

It was mentioned above that electromagnetic radiation of about 21 cm wavelength is received from interstellar atomic hydrogen, and gives us information about the interstellar medium. The explanation of this radiation takes us into several interesting topics, and so will be presented here.

The electron and the proton have angular momentum, called "spin," that is a consequence of relativity. The magnitude of the angular momentum is $\sqrt{[s(s+1)]}(h/2\pi)$, where $s = 1/2$ for both electron and proton (and neutron as well). Relative to any reference direction in space, the spin can take only two orientations, parallel and antiparallel, and the observable momenta are $s(h/2\pi)$, where $s = \pm(1/2)$. Two angular momenta s and s' add to a resultant $S = s + s'$ down to $|s - s'|$. For example, the angular momentum of the deuteron could be 1 or 0; it is actually 1 in the nuclear ground state. The total angular momentum of a nucleus is denoted by I. Hydrogen has $I = 1/2$, deuterium $I = 1$. Nuclei with an odd number of nucleons have half-integral I, while those with an even number have integral I.

Associated with the spin is a magnetic moment. Classically, a particle of mass m and charge q with angular momentum L has a magnetic moment $\mu = (q/2mc)L$. If $q = e$ and $L = h/2\pi$, then $\mu = eh/4\pi mc$. If m is the mass of the electron, then this is the *Bohr magneton*, and happens to be the magnetic moment of the electron, although the spin angular momentum is only $h/4\pi$. Its value is 0.918×10^{-20} erg/gauss. Relativity makes the magnetic moment of spin twice as large as expected. We can write $\mu = g(q/2mc)L$, and say that the *g-factor* g for the electron spin is 2. For orbital motion, it has the classical value 1.

If m is the proton mass, then μ is the *nuclear magneton*, 1836 times smaller than the Bohr magneton. Its value is 5.04929×10^{-24} erg/gauss. The proton magnetic moment is 2.79267 nuclear magnetons, the neutron magnetic moment is

-1.91315 nuclear magnetons. Note that the neutron has a magnetic moment even though its total charge is zero. Nucleons have a complex internal structure as a system of three quarks. Since the proton and the neutron in a deuteron have their spins aligned, the magnetic moment of the deuteron should be the sum of the individual moments, and indeed it is, very closely, at 0.8574 nuclear magnetons.

Because of the spin and orbital motion of the electron, there is a magnetic field at the nucleus. In the ground state of the hydrogen atom, there is no orbital angular momentum, l = 0, so its spectroscopic designation is $^2S_{1/2}$, where the total angular momentum J = 1/2 is due to the electron spin alone. The magnetic field H at the nucleus is aligned with the direction of the electron magnetic moment, which is opposite to the direction of the spin. (Classically, it would seem to be in the opposite direction, but a careful quantum-mechanical investigation shows differently).

The nuclear magnetic moment can assume two orientations with respect to this field, with interaction energies $\pm\mu H$. The total difference in energy between these two states, of parallel and antiparallel spins of electron and nucleus, will be $2\mu H$. These energy differences due to the interaction of the nuclear and electronic spins are called *hyperfine structure*.

In the chill of interstellar space, a hydrogen atom that happens to be in the upper of these two states will eventually jump to the lower state, emitting a photon of frequency f, where $hf = 2\mu H$.

This is so improbable that it is never observed on earth. Transitions between these two states in the laboratory can, however, be stimulated by radiofrequency radiation, and have been easily observed since the development of atomic beam techniques. The transition frequency is 1.4204057 GHz for hydrogen, 0.3273843 GHz for deuterium (because of the smaller nuclear moment). This gives an energy difference for hydrogen of 9.41×10^{-18} erg. From this, we find that the magnetic field at the proton is 33.4T, a surprisingly large value, much larger than macroscopic fields. The 21-cm radiation is caused by a flip of the electron and proton spins from antiparallel to parallel, or from total angular momentum F from 0 to 1.

This discussion of the 21-cm line is closely related to the general phenomenon of magnetic resonance, which is of more general importance, as in modern medical imaging. Let's consider a classical system with angular momentum J and magnetic moment M. These vectors will be parallel, and related by $M = \gamma J$, where γ is the *gyromagnetic ratio*. In the absence of an external magnetic field H, the moments will be directed randomly. An applied magnetic field exerts a torque MxH on the angular momentum that is always perpendicular to it.

Since $dJ/dt = MxH$, the angular momentum, and with it the magnetic moment, rotates about the direction of the magnetic field. Using the proportionality of magnetic moment to angular momentum, we find the equation of motion for the moment to be $dM/dt = \gamma MxH$. In an interval of time dt, the change in the vector moment is $dM = \gamma H\ M \sin\theta$ at right angles to M. If M is rotating with angular velocity ω, $dM = \omega\ M \sin\theta$ as well. Equating these expressions, we hav $\omega = \gamma H$. Since H is a constant vector, we find the remarkable result that all the moments rotate at the same angular velocity, whatever angle they make with the magnetic field. This movement is called *precession*. In an actual system, various mechanisms will exchange energy so that in thermal equilibrium the magnetic moment will eventually point in the direction of H, the state of minimum energy. This is consistent with the expression $E = -\mu \cdot H$ for the energy of a dipole in a magnetic field.

If we now apply an oscillating magnetic field at right angles to the main field, it can be resolved into circularly polarized components that rotate in opposite directions. If the frequency of this rotation is different from the frequency of precession, this field will have no permanent effect. If the frequency is equal to the frequency of precession, a constant torque tending to rotate the moment from the direction of the field will be exerted, and the moments will exchange energy with the oscillating field. This energy exchange can be detected in several ways. By varying the frequency of the oscillating field (or, more easily, by varying the steady magnetic field), the condition of resonance can easily be detected, and the precessional frequency determined.

The equality $\omega = \gamma H$ can be written 2&pif = H(M/J) = $H(4\pi\mu/h)$, where we have used $J = (1/2)(h/2\pi)$, correct for the proton. Then, μ is the proton magnetic moment, 1.410×10 erg/gauss. Now, $f = (2\mu/h)H$. For the proton, $2\mu/h = 4257.52$ Hz/gauss. Therefore, with a steady field of 10,000 gauss (1 T), f = 42.5752 MHz. Resonance will be observed at this frequency for a free proton. In the 21-cm line, we found that the orbital electron produced a field at the proton of 334,000 gauss. In this case, the electron and the nuclear moment are very strongly coupled, and the proton moment will certainly not precess independently. Any practical laboratory field will have very little effect in this case. The expression for the frequency can also be obtained from the transition energy: $hf = 2\mu H$.

In compounds containing hydrogen, the electrons either form closed shells, or occur in pairs in the bonds. In either case, the magnetic field will be closer to zero than to the large value existing in the hydrogen atom. Differences in the field that does exist at different protons will cause differences in the resonance frequency, and the result will be a nuclear resonance *spectrum*, from which much information can be derived.

ORTHO- AND PARAHYDROGEN

Hydrogen has offered many illustrations of surprising quantum-mechanical effects, always confirming them precisely. There are so many that there is no space here to even mention most of them, but one was the discovery of two apparently different kinds of hydrogen gas that could interconvert slowly, called orthohydrogen and parahydrogen. The normal gas was a mixture of 3 parts of the ortho to 1 part of the para. The two forms were chemically and physically identical, except that they differed slightly in spectra and specific heats. This was verified by Bonhφffer and Harteck, and independently by Eucken and Miller, in 1929. If hydrogen was cooled, it eventually became pure parahydrogen (more rapidly in the presence of an activated charcoal catalyst). Alternate lines were missing in the rotational spectrum of parahydrogen, as if only states of even total angular momentum J were present. In normal hydrogen, the lines alternated in intensity with J, those of odd J three times more intense than those of even J.

This classically incomprehensible behaviour was easily explained quantum-mechanically. In a molecule like hydrogen, the two nuclei are identical particles, so the molecule goes over into itself under an (imagined) inversion. Therefore, all its states must be either symmetric or antisymmetric under this transformation, and transitions between symmetric and antisymmetric states are forbidden. Since the two protons have spin 1/2, the total nuclear angular momentum I must be either 1 or 0. For $I = 1$, we have three possibilities, +1, 0, and -1. For $I = 0$, we have only the one. Therefore, in a random mixture, there are three times as many molecules in the nuclear triplet ($I = 1$) state than in the singlet ($I = 0$).

The overall wave function on inversion must be symmetric, since there are four spin-1/2 particles involved. This wave function is the product of the nuclear wave function and the spatial wave functions, including electronic, vibrational and rotational. It turns out that in the ground state, a symmetric nuclear state demands a symmetrical rotational state, and an antisymmetric nuclear state demands an antisymmetrical rotational state. The rotational states are even or odd depending on whether J is even or odd. Therefore, the singlet nuclear state, parahydrogen, (antisymmetric) corresponds to J even, and the triplet nuclear state, orthohydrogen, corresponds to J odd. In cases like this, "ortho" refers to states of greater statistical weight, here 3, and "para" refers to the states of lesser statistical weight.

WATER AND HYDROGEN IONS

The oxygen atom has space for two additional electrons that are easily supplied by two hydrogen atoms. The leftover protons stick to the surface of the oxygen at approximately right angles, though their electrostatic repulsion pushes them farther apart, so that they make an angle of 104° 31'. The O-H bond length is 0.965 Å. This is, of course, the remarkable water molecule, of which a great deal can be said. It has a permanent electric dipole moment, the side with the protons positive, the other end negative. Water molecules tend to stick together when the positive and negative ends approach, forming *hydrogen bonds*. This makes water liquid at ordinary temperatures, and solid below 0°C, when other molecules of similar weight without hydrogen bonds are gaseous (such as NH_3). The polar water molecules, like busy little hands, can pick apart ionic crystals like NaCl by clustering their negative ends around the Na^+ and their positive ends around the Cl^-, freeing enough energy to dissolve the crystal. It is always a balance between the energy released by hydration and the energy released by forming a crystal that determines whether ions dissolve or precipitate.

46 water molecules can form a structure with icosahedral symmetry that possesses six "cages" in which molecules can be trapped. For a discussion of such *clathrates* or gas hydrates.

Liquid water has a density close to 1.000 g/cc, and an index of refraction 1.333. Solid water (ice) has a density close to 0.915, and an index of refraction of 1.309. Water is very important in meteorology, where water vapour is the most effective greenhouse gas.

A molecule like HCl dissolves readily in water, since the hydrations of H^+ and Cl^- are very profitable. Because the concentration of H^+ is large, HCl is called a strong acid, and dissociates nearly completely in water solution. What makes it react like an acid is the hydrogen ion H^+. If there is one mole (6.02×10^{23} ions, or about 1 gram) in one liter of solution, the concentration is called 1M. H^+ concentration, written $[H^+]$, is usually expressed in pH, which is $-\log[H^+]$. For 1M, this gives pH = 0 for a completely dissociated acid. The hydrogen ion is sometimes written H_3O^+ to make the hydration explicit, but this is not really correct, and is irrelevant. One should also then write $H_{2Cl}O^-$.

Water is busy even with itself, picking molecules apart, hydrating the H^+ and OH^-. The latter is called the hydroxyl ion, and is characteristic of bases the way the hydrogen ion is characteristic of acids. This dissociation does not go very far, but in pure water the product of the concentrations $[H^+][OH^-] = 10^{-14}$. This product is constant in equilibrium whatever the concentrations of one or the other ion is by considering reaction kinetics. Since the concentrations of hydrogen ion and hydroxyl ion are equal in this case, $[H^+] = 10^{-7}$M and the pH = 7. If we add one mole of NaOH to make a litre of solution, then $[OH^-]$ = 1M (neglecting the small contribution of the water), so $[H^+]$ is depressed to 10^{-14}M, and the pH = 14. When equal amounts of a 1M acid (pH = 0), contributing $[H^+]$, and a 1M base (pH = 14), contributing $[OH^-]$, are mixed, the ions combine to form water, leaving only enough to make the pH = 7.

Water solutions are a way to achieve chemical reactions easily. The water picks apart the reactants forming hydrated ions, which can then move about freely to find partners. Sodium carbonate and calcium chloride can be mixed as solids, but will not react. When dissolved in water, calcium carbonate is rapidly precipitated because water cannot pick it apart, while sodium and chlorine ions remain as a salt solution. So much chemistry occurs in water solution that other possibilities are often neglected, such as gas reactions or reactions in the fused state, or even reactions on colloidal surfaces, in which atoms can also get together to try different arrangements.

Heavy water, D_2O, has a density of 1.1076 g/cc, melts at 3.813°C, boils at 101.431°C at 1 atm, and has a refractive index of 1.32844. It is said to be harmful to life in large concentrations. Its importance with respect to fission reactors is that unlike ordinary water, D has a very low absorption cross section for thermal neutrons, only 0.00057 barn, compared to 0.33 barn for H. Collisions with D are almost as effective as collisions with H to slow down fast neutrons, so heavy water is still an effective moderator. The O^{16} in both light and heavy water has a very small absorption cross section, less than 20 μbarn. Heavy water is probably the most effective neutron moderator. Using it, reactors can be made that go critical with natural uranium.

PREPARING HYDROGEN

Hydrogen does not occur uncombined on Earth. The only practical source of hydrogen is water. Ultimately, even the hydrogen in hydrocarbons and organic matter came from water. Energy is required to separate the hydrogen from water, and when the hydrogen is subsequently burned, less energy is obtained than was consumed to produce the hydrogen. On the Earth, hydrogen is *not* an energy source, only a means of storing and transporting energy.

The cheapest way to produce hydrogen is as water gas. Coke is burned in air to bring it to red heat. Then the air is shut off and steam is blown into the reactor. The reaction is $C + H_2O \rightarrow CO + H_2$. The water gas, enriched somewhat with hydrocarbons, was supplied as town gas for heating purposes in most cities until natural gas became available. This gas burned with a blue flame, and made putting one's head in the oven an easy suicide. It also was explosive, so that every winter one building or another blew up disastrously. With natural gas, this no longer happens as much, fortunately, but natural gas can also explode, simply not as readily. To make hydrogen, the CO is oxidized to CO_2 catalytically, and then is easily removed by scrubbing with water. This is one of the cheapest ways to make hydrogen, and it will be noticed that not only does is require a good deal more energy than will be recovered by burning the hydrogen, but also produces CO_2.

These days, water gas is little used, and most hydrogen comes from reforming natural gas, usually to make plastics. A typical reaction is $2CH_4 \rightarrow C_2H_4$ (ethylene) + $2H_2$, promoted by a catalyst. The ethylene is a very useful chemical primary, which polymerizes to polyethylene, the plastic PE, among other things. The hydrogen can then be used for hydrogenating unsaturated vegetable oils to turn them into more desirable margarine and Crisco. This is the most probable source of hydrogen for a "hydrogen economy." Again, more energy must be used in making the hydrogen than is recovered by burning it. Incidentally, natural gas is only the most convenient raw material. Hydrogen can be made from other natural organic sources, but much less efficiently, and at the expense of much more energy.

Soon after Volta's discovery of the electric cell, Nicholson and Carlisle electrolyzed water to form O_2 and H_2, then recombined the gases to show that the water was reconstituted. This is an excellent demonstration, showing that two volumes of hydrogen combines with one volume of oxygen to make one volume of water vapour. Many people think of this as a way to manufacture hydrogen, but it is relatively inefficient and requires much more electrical energy than is recovered by burning the product. Nevertheless, photovoltaic cells and wind turbines can produce electricity without any expensive inputs, so this may be a good way to store the energy generated when the sun shines and the wind blows for when it is dark and calm.

The usual laboratory source of hydrogen is the reaction of strong acids with granulated impure zinc. This is an interesting reaction with many applications. The reactions that occur are the discharging of H^+ ions, $2H^+ + 2e^- \rightarrow H_2$, and the oxidation of zinc atoms, $Zn \rightarrow 2e^- + Zn^{++}$. This is really the corrosion of zinc, and similar reactions are responsible for the corrosion of any active metal, such as iron. Adding these two reactions, the net effect is $2H^+ + Zn \rightarrow Zn^{++} + H_2$, where charge and mass balance. How strongly each reaction goes depends logarithmically on the concentration (activity) of the reactants. It is easier to discharge hydrogen from a concentrated solution of H^+ than from a weak one, for example. The tendency of the reactions to proceed is measured by their *electrode potentials*, probably familar to the reader, quoted for standard concentrations. We won't go into the activity dependence here, but it should be kept in mind.

The discharging of hydrogen consumes electrons, so regions where this reaction occurs are sources of electric current (remember that electrons are negative, so an electron moving to the right is a current moving to the left), and have a positive potential. We can allow this to occur at a platinum electrode, which will remain inert. As we draw current from the platinum electrode, bubbles of hydrogen gas will appear on its immersed surface. Whatever voltage is required with respect to the solution to make this happen is conventionally taken as zero. The zinc reaction occurs at a zinc electrode in the solution, and current flows into the electrode as the reaction proceeds. If we connect the platinum and the zinc electrodes, a current flows so long as there is zinc and hydrogen ions. If we apply a reverse voltage between the electrodes, we can eventually make the current stop, and this voltage is the electrode potential of the zinc. At standard concentrations, this voltage turns out to be 0.761 V. In the tables of electrode potentials, the zinc reaction is written as $Zn^{++} + 2e^- \rightarrow Zn$, as a reduction rather than an oxidation, and this makes the tabulated voltage -0.761 V.

The platinum and zinc electrodes in an acid can be used as a battery, of course, and if the H^+ concentration is 1 M, its voltage will be 0.761 V. The platinum electrode will be (+), the zinc (-). At an H^+ concentration of 10^{-7} M, as in pure water, the hydrogen electrode potential is -0.414 V, so the voltage will drop to only 0.347 V. In general, any metal with a more positive reduction potential than hydrogen will displace it from solutions containing H^+. This includes quite a number of metals, notably iron, cobalt, nickel, magnesium, aluminum, titanium, and even lead. However, there are complicating factors, such as the formation of protective surface layers, that may make the release of hydrogen very slow, as in the case of lead, which can be used to hold concentrated sulphuric acid. Aluminium resists acids because of the surface layer of oxide.

Even pure zinc reacts very slowly when immersed in acid. When used in an electrolytic cell, however, pure zinc reacts in proportion to the current flow. Impure zinc immersed in acid, however, reacts rapidly. The reason for this is that the impurities act like the platinum electrode in a cell, becoming positive relative to the zinc, so small cells are formed, and as current flows, hydrogen is evolved at the site of the impurity. In this way, the impurities can serve to corrode all of the zinc, so long as acid is left. In electrical cells, this is called *local action*, and wastes zinc unprofitably. It was traditionally overcome by amalgamating the zinc, which put pure zinc on the surface and short-circuited the parasitic cells. Amalgamated aluminium also makes a good electrode, overcoming the effects of the oxide layer.

Many other reactions produce hydrogen. Sodium reacts rather violently with water to form sodium hydroxide and hydrogen: $Na + H_2O \rightarrow NaOH + (1/2)H_2$. Sodium hydride reacts similarly: $NaH + H_2O \rightarrow NaOH + H_2$. Zinc also dissolves in strong alkalis to release hydrogen: $Zn + 2NaOH \rightarrow Na_2ZnO_2 + H_2$. Methanol can be catalytically oxidized to give hydrogen: $CH_3OH + H_2O \rightarrow CO_2 + 3H_2$. This reaction is specially useful for portable hydrogen generators.

USING HYDROGEN

The principal use of hydrogen today is not as a fuel, but as a chemical feedstock. It is produced as a by-product of plastics manufacture, and is then used in a variety of ways. One is for the hydrogenation of vegetable oils, as mentioned above. Another is the production of methanol by the Patart synthesis, $2H_2 + CO \rightarrow CH_3OH$, with the aid of a catalyst. This is the inverse of the hydrogen-producing reaction of the preceding paragraph.

Most of the proposed uses of hydrogen, however, produce heat by burning it in oxygen. This may be done in an open flame, in an internal-combustion engine, in a gas turbine, or fuel cell. The basic reaction is $H_2 + (1/2)O_2 \rightarrow H_2O + Q$. If the water is condensed to liquid at a standard state, Q = 68.32 kcal (per mol). If the water is exhausted as a vapour, then Q = 57.80 kcal. The difference is the heat of vaporization of water. These heats may also be expressed as 34.18 kcal/g or 29.15 kcal/g. Multiplying kcal/g by 1800 gives Btu/lb. This is a very high energy content on a weight basis, but unfortunately hydrogen is light. On a volume basis, it gives only 325 Btu/cuft, while methane gives 1008 Btu/cuft.

Liquid hydrogen and liquid oxygen make a compact, high-energy rocket fuel, but handling and storage of the liquid gases is very difficult, and is not practical for other vehicles.

Because of the energy required to liquefy the hydrogen, it is rather inefficient as well. A commercial steel cylinder of compressed hydrogen contains 176 cuft at STP, less than a pound, which will give only 57,200 Btu in all. The pressure in such a cylinder is 120-150 atm (1800-2200 psi). The problem of hydrogen storage is a severe one. Unlike for natural gas, liquefaction is not an option for widespread use.

Iron, gold, platinum and palladium adsorb hydrogen strongly. Although the hydrogen diffuses rapidly through these metals, the adsorption is a surface effect and increases when the metals are in colloidal form. One volume of colloidal Pd will adsorb 870 volumes of hydrogen, releasing it when heated. Apparently, this is in terms of hydrogen at STP, and the amount adsorbed does not increase proportionally to the pressure. Relative to steel tanks, Pd only decreases the volume required by a factor of about 5.8. A 13-gallon fuel tank filled with gasoline (87 lb) will yield about 1.8×10^6 Btu. 30 lb of hydrogen would give the same yield, so 32 steel tanks of compressed gas would be required for the same range. These tanks could be replaced by 5.5 times the volume of one tank in palladium, which would be very costly, not to mention heavy. The use of hydrogen in vehicles will be severely restricted by these considerations, which are similar to those of vehicles using storage batteries, but worse. Fuel cells have no bearing on the problem. Hydrogen filling stations had better be closely spaced.

Hydrogen diffuses more rapidly than any other gas. It will even diffuse through metals, especially those that adsorb it, as well as through quartz, and probably also glass. It will search out the smallest leak, especially in valves and similar devices, which must be carefully designed to hold hydrogen. This property may well result in occasional surprises for hydrogen users.

Hydrogen can be burned in a torch with air or oxygen. An air-hydrogen torch flame reaches 2045°C, while an oxyhydrogen flame reaches 2660°C. Flame temperatures are subject to considerable uncertainty, and depend on the mixture used. The hydrogen flame contains no carbon, and so is invisible. A mixture of 45% 1025 Btu/cuft natural gas and oxygen produces a flame temperature of 2930°C, it is reported. Oxyhydrogen flames playing on CaO refractories produced the "limelight" used in theatrical productions before electrical lighting.

Hydrogen can be dissociated in an electrical arc in an atomic hydrogen torch. The recombination of the atoms produces a very high temperature, 4000°C to 5000°C, that can be used for welding. In addition, the atmosphere surrounding the weld is nonoxidizing, so flux is not required. The reaction is $2H \rightarrow H_2 + 102.6$ kcal. It is not necessary to burn the hydrogen, but this adds extra heat and eliminates an explosion hazard.

I have recently heard of a microorganism, carboxydothermus hydrogenoformous, that consumes carbon monoxide and water, releasing hydrogen and carbon dioxide. This could be another source of hydrogen, but note that carbon dioxide is also produced. This is just the water gas reaction, but mediated by life.

Lighter-than-Air Craft

Hydrogen was famously used to fill lighter-than-air craft, since it is the lightest gas available. At 0°C and 1 atm, air weighs 1.293 kg/m^3, a far from negligible weight, though not apparent to us living immersed in it. This is the buoyant force available for lifting. The essential thing is making the weight of a craft less than the available buoyancy; that is, the average density of the craft must be less than that of air. This is achieved by filling most of the volume with something lighter than air. Air has a molecular weight of 28.966 g/mol. Any gas with a smaller molecular weight will provide lift. The possibilities are hot air, water vapour (M = 18), ammonia (M = 17), methane (M = 16), helium (M = 4), and of course, hydrogen (M = 2). Since one gram mol of gas at 0°C and 1 atm occupies about 22.4 l, it is easy to calculate the net buoyant force in any case.

Hydrogen weighs 89.3 g/m^3, so the maximum lifting force is 1293 - 89 = 1204 g/m^3. Helium weighs twice as much, so the lifting force with helium is 1114 g/^{3}, only 7.4% less. Hydrogen and helium are, therefore, practically equivalent in lifting force. At the other end of the scale, air heated to 100°C weighs 946 g/m^3, providing only 258 g/m^3 lifting force. The gas container must be as light as possible, so it is generally a thin-skinned bag. A number of bags may be used so that all the buoyancy is not lost if a bag is ruptured.

Lighter-than-air craft may be classified as tethered balloons, free balloons, and dirigibles or airships. The dirigibles may be classified as nonrigid or rigid. Nonrigid airships, with no internal frame defining its shape, are often called *blimps*, from the British code name for such (limp) ships under development, which were termed A-limp, B-limp and so on. Rigid airships are often called after their inventor, Zeppelin.

The Montgolfier brothers arranged the first balloon ascents in 1783, using hot air and a fabric envelope. In the same year, Charles demonstrated hydrogen-filled balloons. In 1836, the Great Balloon of Nassau, 85,000 cuft and hydrogen-filled (total lift 2890 kg) made a flight from England to Germany.

Balloons were used for observation in the Franco-Prussian War of 1870, during which Lιon Gambetta made his celebrated escape from beseiged Paris by balloon. Manned balloon ascents reached 113,700 ft (34,700 m) in 1961, and an unmanned balloon reached 168,000 ft (51,200 m). The largest scientific balloon had a capacity of 26,000,000 cuft. Sport ballooning is still practiced, with hydrogen in Europe and hot air in the United States, provided by a butane or propane burner. Long-distance balloon flights are made as stunts. These free balloons are at the mercy of the winds, so they generally travel from west to east, and cannot make return trips.

The first dirigibles awaited the availability of engines of low weight but sufficient power. Graf Ferdinand von Zeppelin (1838-1917) brought out the first dirigible in 1900, hydrogen-filled with an aluminium frame. Its volume was 399,000 cuft (11,300 m^3), so its gross lift was 13,600 kg. The application of

Zeppelins as terror weapons was quickly perceived, and they were used for rather ineffective bombing of civilians.

The first Zeppelin raid against Britian was on 19 January 1915. Of the 67 Zeppelins made during the War, only 16 survived, since they were very susceptible to attack. After the War, military authorities in Britain, the United States and Italy became interested in the possibilities of dirigibles, especially in view of the possibility of helium filling, due to the anticipated availability of helium from the United States.

The British R34 of 1919, 1,980,000 cuft, made the first transatlantic crossing, but was lost in 1921. The *Roma*, supplied by Italy to the United States in 1921, 1,200,000 cuft, was lost in 1922. The U.S. dirigible *Shenandoah* (originally ZR1), 2,115,000 cuft, built in 1923, was the first filled with helium.

It was lost in 1925 at the cost of 14 lives. The *Los Angeles*, 2,475,000 cu. ft., followed in 1924. The *Graf Zeppelin*, 3,710,000 cu. ft., of 1928, entered regular long-distance passenger service.

These ships were powered mostly by multiple marine diesel engines. In 1929, Britain built the large R100 and R101, of 5,000,000 cu. ft..

When R101 was lost at Beauvais in 1930 with 46 killed, Britain ceased the development of airships. In 1931 and 1933, the United States constructed the *Akron* and *Macon* (they were named like navy cruisers), 6,500,000 cu. ft.. They were wrecked in 1933 and 1935, respectively, and the United States terminated the development of rigid airships. Blimps, for submarine warfare and air defence, were used by the U.S. Navy until 1983. A few blimps are still maintained for advertising purposes.

It is clear that all these large ships had very short lives, being unable to cope with the winds. This is a general characteristic, and every new generation of airship enthusiasts has to learn it anew, it appears. No more dangerous and unreliable method of transportation than the airship was ever devised, but its charm is undeniable. The *von Hindenburg*, a hydrogen-filled airship of 6,710,000 cu. ft., was built in 1936 for transatlantic service. It very famously burned at Lakehurst, NJ on 6 May 1937 with the loss of 36 passengers out of a total of 92. As disasters go, this was pretty tame compared with later ones, in which hundreds die in numerous air crashes. Nevertheless, it was a very impressive disaster. The flames in the images are from the fabric bags; the hydrogen flame is invisible.

The accident was probably due to a small spark combined with leakage of air into the hydrogen. Although much is made of the danger of hydrogen, this was the only significant accident due to this cause. Consideration of the fates of the previous dirigibles show that hydrogen fires were unimportant. In military connections, however, they are more significant.

Any large balloon or airship must carry ballast, either water or sand, to adjust the buoyancy as the lifting gas diffuses away, or is heated or cooled. Discarding ballast causes the ship to rise, venting lifting gas causes it to descend. The excess of the lift over the weight of the craft is called the *free lift*. For a small balloon, it can be measured by a scale before the balloon is released. The amount of free lift determines how fast the balloon will rise, and how high it will ascend.

Free balloons are of three general types: (1) rubber or neoprene elastic balloons; (2) polyethylene or similar inelastic balloons; and (3) superpressure balloons of constant lift. The neoprene balloon, typically used for radiosondes, expands as the balloon rises until the balloon bursts, and the payload descends on a parachute.

A polyethylene balloon, for lifting heavier loads, is only partially filled at the surface, becoming spherical at altitude. A valve dumps lifting gas to maintain the altitude constant. At nightfall, the gas cools and the balloon descends to the surface.

The superpressure balloon adjusts the tension in its skin to suit changing conditions, and remains at its equilibrium altitude until the lifting gas diffuses away.

Barrage balloons, tethered balloons looking like blimps, but unpowered, were used in air defence in World War II. Low-flying aircraft would fly into wires suspended below them. Higher-flying aircraft were a better target. Of course, they are now completely obsolete.

Filling of such balloons with helium, as is done in the United States, is an arrogant waste of a scarce and valuable resource. Only the hazards of the hydrogen supply for filling are avoided by the use of helium, as the loss of a released balloon by fire (which does not happen anyway) is inconsequential.

Hydrogen for filling balloons requires a source of a size between that of an industrial supply, and a laboratory supply. Some suitable methods were mentioned above. The apparatus can be put easily into a road trailer, and moved to wherever it is needed. These hydrogen sources have been confused with weapons of mass destruction in some cases by the desperate and unintelligent.

Chapter 2

Helium

This article is about the extraordinary gas helium, but at the same time it will be convenient to talk about the other members of its family in the periodic table. They are known collectively as the noble, inert or rare gases, because they do not form chemical compounds. They also are quite rare, and became known only at the end of the 19th century. An especially remarkable fact about helium is that it is available in large, industrial quantities, and was even used to fill airships. The availability of helium in such large quantities is an almost unbelievable surprise. One purpose of this article is to make clear how we get our helium.

In 1785, Henry Cavendish found that about 1/120 of the volume of air was a gas that was not oxygen, nitrogen or carbon dioxide, and which would not react chemically. This discovery, like others of Cavendish's, was largely forgotten. Indeed, Cavendish himself thought it might just be experimental error. On the 18th August 1868 there was a total eclipse of the sun visible from India.

Pierre Janssen studied the chromosphere of the sun, which was until then only visible during total eclipses, with a spectroscope and found its spectrum to consist of bright lines, as are emitted by excited gases. One very prominent yellow line at 587.56 nm did not correspond to any known element. At the same time, Norman Lockyer finally perfected his spectroheliograph, which allowed the study of the chromosphere at any time, and he, too observed the mysterious yellow line. Lockyer made the leap to propose that this line corresponded to some element common on the sun, but rare on earth, and he called that supposed element *helium*. Helium does produce a Fraunhofer line in the solar spectrum, D_3, but it is too close to the sodium D lines to be easily observed. A search for helium began, but nobody knew where to look.

The search for terrestrial helium was unavailing, and the source of the yellow line remained obscure. Then, after 1890, Rayleigh and Ramsay discovered that nitrogen produced chemically (by heating NH_4NO_3, for example) was lighter than nitrogen separated from air by fractional distillation of liquid air. Rayleigh recalled Cavendish's forgotten observation, and it set him thinking. Liquefaction of gases was a new science then, in which Kamerlingh Onnes was a pioneer. It turned out to help greatly in the search for helium. Ramsay separated the heavier component from atmospheric nitrogen in 1894, calling it *argon*. This name comes from Greek a-ergos (in which the "e" was elided, making the word identical to another Greek word meaning "brilliant"). This means "not-working" or lazy, inert or inactive, which perfectly describes argon's chemical interests. The -on is simply the neuter ending.

Ramsay heard of inert gases given off by uranium minerals in the studies of W. F. Hillebrand of the USGS published in 1891. The gas given off when uraninite was dissolved in acid was studied spectroscopically by Hillebrand and found to be apparently only nitrogen. The nitrogen spectrum had obscured the feeble yellow line from the small amount of helium present. Ramsay obtained some clevite, another uranium mineral that by chance was available, and treated it the same way. Taking care to eliminate any nitrogen impurity, he saw a strange spectrum, and named the gas he had found *crypton*. Some was sent to William Crookes, who identified the gas as the long-sought helium. Therefore, in 1895, terrestrial helium was finally found.

With argon and helium now identified, the periodic table suggested that there were other members of the family. Ramsay, with the help of Travers, went to work to see if they could be discovered in the argon separated from atmospheric nitrogen, by fractional distillation of the liquefied gases. In

1898, they announced the discovery of neon, krypton and xenon. Kayser found that helium was also present in air. Therefore, all of the gases were found in atmospheric air. Argon was present to about 1% by volume, a considerable amount, but the other gases were by far scarcer. In ppm by volume, air has the composition Ar, 9300; CO_2, 300; Ne, 18; He, 5.2; Kr, 1.0; Xe, 0.8. Air is, therefore, a good source of argon in large volumes, but Ne, He, Kr and Xe can be separated from it only in small amounts and at high cost. These amounts would be sufficient for discharge tubes, which use only very small amounts, and for scientific purposes, but not for any wide-scale use.

In 1907, Rutherford showed that the alpha-particles emitted at high velocity from radioactive substances were helium nuclei. When alpha particles were allowed to enter a vacuum through a thin screen, the characteristic spectrum of helium was soon observed when an electrical discharge was produced. Most of the helium on the earth, in fact, was created in the form of alpha particles. The disintegration of radioactive substances has been going on throughout the existence of the earth.

This radioactivity occurs mainly in the upper mantle, just beneath the crust, where the heavy atoms have been concentrated by convection. Helium that reaches the atmosphere is "boiled off" by the earth and is dissipated into space, just as hydrogen and other light gases are. Water, with a molecular weight of 18, is just heavy enough to have been largely retained (but it seems to have boiled off of smaller Mars already). Neon, with an atomic weight of 20, has been retained by gravity, as have, of course, the heavier xenon and krypton. On the other hand, the original abundance of these atoms on the earth decreases rather sharply as the atomic weight increases. The abundance of the noble gases in the atmosphere seems to be well accounted for by these considerations. Argon, it should be noted, is the product of the beta-decay of naturally-occurring K^{40}, which may be the source of most of it. Although K^{40} is rare (1 part in 10^4), there is a lot of natural potassium.

The Source of Helium

The helium from most of the alpha particles, however, has been trapped deep within the earth, not far from the top of the mantle. Helium diffuses very rapidly, so it can be expected to have been mobile in the crust. The Great Plains of the United States consist of a thin crust of sedimentary rock on a basement of igneous rock that would be a source of radiogenic helium. The helium has collected where other gases have collected, depending on the chances of distribution. In some of these gases, helium is present up to about 8% by volume, a surprising amount. However, in most gases the abundance is less than 1%, sometimes much less, and the occurrence is very rare on a world-wide basis. Some natural gas or well helium, as radiogenic helium from the earth is also called, is found in the Canadian west, and, in lesser quantities, in gas associated with the blue rock salt of Germany, and a few other locations. There may also be helium in Siberia, where the geologic conditions resemble those of the Great Plains.

Porous and permeable rocks deep below the surface are often saturated with salt water, at a pressure of the same order as the hydrostatic pressure at that depth, called the *formation pressure*. Since 33 ft of water corresponds to 1 atm, or 14.7 psi, the formation pressure at 3300 ft would already be 1470 psi. Gases in permeable rocks collect at the highest points they can reach, and there form a "cap" of gas, possibly underlaid with liquid petroleum. A well drilled into the part of the formation containing liquid petroleum would soon fill with the liquid and flow out of the hole, as quickly as it could ooze into the hole. The weight of the petroleum would largely balance the formation pressure, so the oil would generally flow placidly. In rare cases, the formation pressure might be high enough to project a jet into the air, a "gusher." As formation pressure is lost, as it always is when holes are drilled, the oil might not reach the surface, and then it would have to be pumped out mechanically. Oil cannot be sucked from rocks; you can only collect what shows up.

With gas, what happens is quite different. The gas is so light that it cannot balance the formation pressure, and the gas exits from the hole at high pressure, unless confined (perhaps by the fluids used in drilling the well). Such a well was called a "blower" in early days, and the owners hardly realized that what was blowing out was, in effect, their wealth, since with depleted formation pressure little oil or gas could be produced. A gas well is exhausted when the pressure is no longer sufficient to bring the gas to the surface at sufficient pressure. These days, much gas in the western United States is produced from shallow coal-bed wells or from formations of small permeability, or "tight" formations. This gas has to be compressed before it can be put into the gas pipeline.

Natural gas is often thought of as methane, CH_4, and most marketable natural gas is chiefly methane. However, the gas ranges from a rich gas containing considerable ethylene and even heavier components, which give it a high heating value, down to lean gases containg so much nitrogen that it will not even burn if ignited. The bacteria that produced the oil also leave behind carbon dioxide and hydrogen sulphide, which are often found in the gas. Hydrogen sulphide makes the gas poisonous and of foul odor; such gas is called "sour." The hydrogen sulphide is relatively easily removed, and the sulphur that results is a valuable by-product. Some gas is almost pure carbon dioxide, and many wells are produced for the carbon dioxide, a valuable chemical. Indeed, carbon dioxide is often pumped underground to repressurize a formation, and serve as a lubricant for getting the oil out. Sweet natural gas has little odor and is not poisonous.

Some of this natural gas also contains helium. Gas rich in helium usually also has considerable nitrogen, but much nitrogen does not imply helium. Helium was first found in a "blower" at Dexter, Kansas southeast of Wichita near Winfield in 1905. This gas would not burn, to the disappointment of the locals, but it did contain 1.84% helium, as H. P. Cady at Kansas University found. This was a chemical curiosity, since it had been expected that helium would be very rare, but helium had no uses at the time. This status changed completely during World War I with the introduction of Zeppelins.

These rigid airships were filled with hydrogen, as indeed this was the only thing available other than hot air. Hydrogen diffused out of an inflated airship, while oxygen from the air diffused in. Not only was buoyancy lost, but more dangerously, an explosive mixture could be formed. When a light is touched to a balloon filled with pure hydrogen, it burns rapidly. When a light is touched to a balloon filled with an explosive mixture of hydrogen and oxygen (and the mixture is explosive over a wide range of concentrations) there is a violent catastrophe. In the well-known Hindenburg disaster, the hydrogen merely burned. If the mixture had been contaminated with oxygen, everything would have been blown to smithereens.

The lifting power of hydrogen is proportional to the difference in molecular weights between hydrogen and air, which is 29 - 2 = 27. The lifting power of helium is 29 - 4 = 25, which is 93% that of hydrogen. Replacing hydrogen with helium sacrifices little lifting power, but gains the immense advantage of not being flammable. After World War I, the great advantages of helium-filled dirigibles and barrage balloons were evident, so there was great military interest in helium. Without the Kansas helium, this would have been utterly impossible. Therefore, the development of helium became something of a military secret. Secrets are one of the pleasures of a military life, and they were enjoyed to the full over helium. Seibel tells the story of the development of helium for military purposes by the United States.

In Kansas, a thin layer of mainly Pennsylvanian rocks, in which there was enough organic activity to produce natural gas, probably by bacterial action, overlies granitic rocks of the continental shield. In fact, the Nemaha Ridge is a granite mountain trending north-south through central Kansas beneath the sediments, and apparently giving off helium as a result of its radioactivity. This midcontinent field was rich in crude oil, and especially in natural gas. Because natural gas was difficult to transport economically (it is too light), it was mostly considered a waste product at the remote locations of oil wells, and what could not be used locally was flared off.

However, high pressure gas pipelines were economical where there were large quantities of gas and concentrated markets for that gas.

After the First World War, the great natural gas fields of southwestern Kansas, the Texas and Oklahoma panhandles and New Mexico were just beginning to be developed. Fort Worth, however, already had natural gas, and that coming from the Petrolia field was rich in helium. This gas served to develop the methods for separation of helium devised by the Bureau of Mines. There are two general ways to do this. One is to adsorb all but the helium on cooled activated charcoal, and the other is to liquefy the gas and separate the helium by fractional distillation. The second method proved the most economical, but the first was used to some extent to purify the gas obtained when airships were refilled. Just as in the case of hydrogen, air diffused into the helium and had to be removed to preserve the lifting power, but, fortunately, not to avoid an explosive mixture.

Helium at room temperature is quite an ordinary gas, but its small atomic weight means that it diffuses very rapidly, even through very small apertures. Pumps made for other gases proved very inefficient with helium, until special valve designs were devised. Hydrogen, of course, behaves the same way, but diffuses even more rapidly. Even barrage balloons are now obsolete, so that lighter-than-air craft are no longer of military interest. However, the other military uses of helium have become, if anything, of even greater importance.

Eventually there were nine helium conservation plants in Kansas and the panhandles, connected by a helium pipline from Bushton, in central Kansas, to the Cliffside field just north of Amarillo, with several branches. Helium was stored successfully underground in the Cliffside field as the field itself became depleted. Of course, the problem in helium storage is preventing its leakage. The grade A helium now produced is 99.995% pure, a remarkable achievement.

Gas with 8% helium was discovered near Thatcher, Colorado. However, the formation pressure was low, and the carbon dioxide content was 14%. The Rattlesnake field near

Shiprock, New Mexico also had 8% helium, and was nonflammable, so the residual gas could simply be vented. A large helium plant was built at Shiprock, in spite of its distance from the railway, and that water had to be obtained at some distance from the San Juan River. Fuel gas had to be furnished from gas wells 35 miles away, that had gas that would burn. The first well, Navajo No. 1, flowed 30 million cubic feet per day at a wellhead pressure of 2900 psi. The plant went into production in 1944. The gas is now exhausted, and the plant dismantled. This plant is a good example of the limited life of gas fields.

The natural gas fields of the midcontinent are now all nearing exhaustion, and with them will go our principal source of helium. So much helium has been produced in the past that no shortage is envisioned by the authorities, but it is sure to come someday. Helium occurs only exceptionally in natural gas, and the U.S. was fortunate in its abundance. However, like crude oil, iron ore and other finite national resources, it is nearing exhaustion. There is no other source of helium, and no substitute for it.

WHY THE INERT GASES ARE INERT

The reason why the inert gases are inert can be quite clearly explained. It is not just haughtiness on their part, but a consequence of quantum mechanics. All atoms are just a very small, massive, nucleus of electrical charge +Ze, that has attracted Z electrons with total charge -Ze to make an electrically neutral assembly. The electrons are accommodated in discrete states that occupy definite volumes. If we start with a nucleus of charge Z and build up the atom by adding electrons one by one, the first electron can occupy a state, or *orbital*, close to the nucleus, designated 1s.

The second electron can also occupy this state, provided that its spin in opposite to that of the first electron. The third electron must go into a state outside of this, and so of less energy, to a state designated by 2s, or to one of the three states designated 2p. Eight electrons can be accommodated in this "shell," after which we must go to states of still higher energy.

When we have added Z electrons in this way, the atom is neutral and can form part of macroscopic matter, which cannot be electrically charged because of the strong forces that would result if it were not. If we put the electrons in states of the lowest possible energy, we have the *ground state* of the atom. Any other distribution gives us an *excited state* of higher energy. An atom in an excited state will emit electromagnetic radiation and fall into the ground state in a short time.

We wish to compare the behaviour of hydrogen, Z = 1, with a nucleus consisting of a proton, and helium, Z = 2, with a nucleus consisting of two protons and two neutrons, strongly held together by nuclear forces. The single electron of hydrogen in its lowest state is in the 1s orbital, from which 13.5 eV are required to eject it. The two electrons in helium are in the 1s orbital with opposite spins, which is the *configuration* $(1s)^2$. The next electron would have to go into an orbital of considerably higher energy (in lithium), so the 1-, or K-shell is said to be "filled."

If we bring two hydrogen atoms near to one another, the electrons can arrange themselves in different ways, or molecular orbitals, around the two protons. If we imagine the molecular orbital as a linear combination of the atomic orbitals, we can form the orbitals 1s + 1s and 1s - 1s, adding in one case and subtracting in the other. When we normalize the result (make its integral over all space equal to 1) we get two orbitals that are approximately those of the combination of the two atoms. These are designated (σ_g1s) for the sum, and (σ_u1s) for the difference. The absolute value squared of these wave functions indicate the average electron density. In the first state, it is easy to see that the electron is mainly between the nuclei, but in the second it is beyond them. Calculation shows that when the electron is between the nuclei, the shielding of the repulsion of the two nuclei is complete enough that the energy is lower than that of the two bare nuclei, which then attract rather than repel. Such an orbital is called *bonding*. In the other state, the nuclei repel strongly, and the orbital is called *antibonding*. For one 1s state on each atom, we get two molecular orbitals.

When two hydrogen atoms are brought near to one another, there are two electrons to dispose of, and both can be placed in the bonding 1s molecular orbital. The electrons are mainly between the protons, neutralizing their charge exactly, and causing the protons to attract one another. A potential well is produced that is deepest at some finite separation of the protons. The two atoms then hang together quite strongly, and can vibrate and rotate if thermally agitated. A stable, neutral diatomic molecule has been formed, H_2. 4.476 eV is required to break it apart, and this is much greater than the normal thermal energy.

If we do this with helium atoms, the first two electrons will fit into the bonding orbital, as in hydrogen. The third and fourth electrons, however, will have to go into the antibonding orbital, and as a result the atoms will not attract one another enough to stick together; their mutual potential energy curve will have no minimum. Therefore, there is no stable He_2 molecule in the ground state. However, if we use the 2s orbitals, then we can form a bonding orbital ($\sigma_g 2s$) which will be more favorable than the antibonding 1s orbital. If we put the fourth electron in such a state, then there is sufficient attraction to hold the molecule together, and we have $He_2^*(\sigma_g 1s)^2(\sigma_u 1s)(\sigma_g 2s)$. The * indicates that the molecule is in an excited state. Obviously, there may be other possibilities, but this is one structure that is observed, and whose spectrum has been investigated.

Another possibility is to ionize one of the atoms, so that there are only three electrons, which can be accommodated so that the atoms stick together. He_2^+, like He_2^*, exists and has been studied. Since it is electrically charged, it can occurr only in small amounts and cannot make any bulk matter, because of the strong electrical forces. He_2^+ will attract an electron, become He_2^*, which in turn will radiate (emit photons), reach the ground state He_2, and fall apart. This is the reason helium is an inert gas, and will not combine with other atoms, just as it will not combine with itself, to form a stable diatomic molecule. Some success has been experienced in forming compounds of inert gases with very strong electron acceptors such as fluorine, but any such compounds are very fragile, and the statement that the inert gases form no chemical bonds is a very good approximation. Each inert gas has a completed outer shell of 8 electrons (configuration s^2p^6), and the jump to the next available orbitals is a large one, so all the inert gases act the same way.

This is yet another example that atoms or elements have no innate properties. Everything is determined by structure. The visible universe is held together by electrical forces, and governed by quantum mechanics. This explanation of matter has been completely successful, and is of universal application.

The ground state of the helium atom has zero spin, as we have seen, because there is only one orbital available, and two electrons can occupy it only if their spins are opposite, so that the angular momentum adds to zero. If the two electrons are in different orbitals, then their spins may be either opposite and give singlet states, or may be parallel and add to 1, giving triplet states, so called because three different states can be formed that have the same energy in the absence of a magnetic field. For a singlet, only one such state is possible. Any spectral line is the result of a transition from one stationary state to another of the atom. The selection rule that singlets may not combine with triplets is very strictly observed in helium. The spectrum, therefore, seems to be separate spectra of singlet helium, called parhelium, and of triplet helium, called orthohelium, as if these were two different gases mixed in the ratio of 1:3. Of course, this is not really the case, but the strength of the selection rule is brought out. One result is that the triplet state of the configuration 1s2s, the lowest excited configuration, cannot combine with the ground state, and so is called metastable. Any helium that winds up in this state will remain there, unless dislodged by a collision or similar disturbance. The observed helium spectrum is mainly due to excitation of one electron, the other remaining in the 1s state.

The helium atom was an excellent test of the new theory of wave mechanics. The earlier orbit theory of Bohr and Sommerfeld failed completely with hydrogen, but the new mechanics gave results more accurate than the current experimental values for the ground state energy of helium. The first ionization potential of helium is 24.580 V, and the second is 54.403 V, so the ground state is 78.983 eV below the alpha particle and two free electrons. Recall that the ionization potential of hydrogen is 13.5 V.

THE INERT GASES AS GASES

We note that the hydrogen molecule is really rather inert itself. If it encounters another hydrogen molecule, the situation for the electrons is much as in the case of two helium atoms. There is not enough room to accommodate the four electrons between the nuclear charges, so there is very little attraction. The other diatomic atmospheric gases, oxygen and nitrogen, behave the same way. An oxygen atom is a very active thing, and will combine with about any other available atom.

However, the diatomic molecule is practically inert. Nitrogen is similar, but diatomic molecules of oxygen and nitrogen can bumble around in the air with nothing happening. It takes a lightning stroke, which can break up a diatomic molecule, to encourage the formation of only a little nitrogen oxides. The atmosphere is quite safe from becoming nitrogen oxides in a grand conflagration. Similarly, oxygen and hydrogen, as diatomic molecules, can coexist peacefully, at least in the dark. The slightest spark that tears apart a diatomic molecule can release a chain reaction that proceeds explosively. The difference between these diatomic molecules and the inert gases is simply that the diatomic molecules can be torn apart with sufficient violence, while the inert gas atoms cannot be.

Actually, two helium atoms do attract one another slightly because of the interaction of their fluctuating electrical fields, that act like rapidly moving dipoles averaging to zero. These are the London dispersion forces, or Van der Waals forces. These forces become stronger as the number of electrons increases, so that Xenon atoms are relatively sticky. The main force between two helium atoms is a strong repulsion at short distances, so that they resemble hard spheres. Helium is a close approximation to the hard-sphere gas studied in kinetic theory, and so is a good example of the theory. At a temperature T, the average kinetic energy per molecule is 3kT/2, and this is the only internal energy of the gas to a good approximation. This means that the molar specific heat at constant volume is C_v = 3R/2, the molar specific heat at constant pressure is C_p = 5R/2, the ratio of specific heats $\gamma = 5/3$, and the gas obeys the ideal gas law pV = RT, where V is the molar volume, and R is the molar gas constant. The speed of sound c is given by $c^2 = \gamma RT/M$. Helium obeys all these relations very well.

Actually, the finite volume of the atoms and the slight attraction between them affect the equation of state, the relation between p, V and T. The van der Waals equation of state, $(p + a/V^2)(V - b) = RT$, accounts for these effects through the constants a and b, different for each gas. If p is in atmospheres, V in litre/mol and T in K, then R = 0.08217. For helium, a = 0.03412 and b = 0.02370. These values are remarkably small, less than for any other gas, and show how close helium is to an ideal gas. When we get to xenon, a = 4.194 and b = 0.05105. The larger, more usual, value of a reflects the greater attraction due to dispersion forces, but the value of b shows that the atoms are still small. The atomic radii, in Å of the inert gases are: He, 0.53; Ne, 1.60; A, 1.91; Kr, 1.98; Xe, 2.18. The diameters range from 1 to 4 Å, typical atomic diameters, small compared to most molecules.

The speed of sound in helium at 300K is 1010 m/s, about three times greater than in air at the same temperature, where the speed is 347 m/s. Acoustic resonators filled with helium will resonate at a higher frequency than those same resonators filled with air. The pitch will be about a twelfth higher, an octave and a half. The same is true for hydrogen, where the speed of sound is 1316 m/s, but the pitch is raised even more, close to two octaves. Professor Tyndall used to amaze the children at the annual Royal Institution lectures by breathing in hydrogen, and then speaking with the unexpectedly squeaky voice that resulted. This, of course, is extremely dangerous and Tyndall was lucky. It is much more safely done with helium (now that helium is available cheaply in the required amounts), and is a familiar party trick. Breathing xenon the same way should make anyone a booming bass, but the trick would be a costly one.

Breathing helium (or any other inert gas except radon) is quite safe, so long as enough oxygen is supplied to avoid asphyxiation. In fact, 80% He, 20% O_2 is a good breathing mixture for divers and others working under pressure (with the oxygen content properly regulated). It happens that helium is the gas least soluble in water, only 0.97 parts in 100 parts at 0°C. Nitrogen is soluble to the extent of 2.35 parts. When air is breathed under pressure, nitrogen dissolves in the blood. When the pressure is relieved, the nitrogen comes out of solution as bubbles, which create havoc in the circulatory system, causing the "bends" or "caisson disease." Helium dissolves to a much lesser extent, and decompression is quicker. Helium is also used in breathing mixtures for asthmatics, since its rapid rate of diffusion opens up congested lungs more easily than ordinary air does. In anesthesia, helium is also used as a carrier gas.

The mass of helium on the O = 16 scale is 4.00387 amu. The mass of two hydrogen atoms and two neutrons is 4.03426 amu. The difference, 0.03039 amu, corresponds to the binding energy of the alpha particle, or helium nucleus, about 28.3 MeV. This demonstrates the great strength of nuclear forces, as well as the stability of the alpha particle. Helium is produced in large amounts by thermonuclear reactions in the sun. Hydrogen and helium are, by far, the most common elements in the universe, but only small amounts of them appear on the earth.

Chapter 3

Xenon and Radon

These are the heaviest of the inert gases, and have some peculiarities. Xenon is, amazingly, an anaesthetic, in spite of the fact that it is inert. This is not completely explained, but is supposed to be associated with the strange behaviour of xenon and water. 46 water molecules can hydrogen-bond themselves into a complex molecule with icosahedral symmetry that forms eight "cages" of the right size to accommodate xenon atoms, stabilized by dispersion forces between the xenon and the water. This $8Xe{\cdot}46H_2O$ is called a *clathrate*, from the Greek for a cage. Eight methane molecules can take the place of eight xenon atoms, to form an ice-like methane hydrate that can block pipe lines even at temperatures well above 0°C. Possibly the xenon encourages the formation of such clathrates in the brain, and the other smaller cages that are formed can clathrate ions important in brain function. For more information on gas hydrates.

Radon is the heaviest inert gas, and all its 15 isotopes, ranging from mass 206 to 222, are radioactive. The one with the longest half-life, Rn^{222}, was the one originally called radon, or *radium emanation*. It is produced by the alpha decay of ordinary radium, Ra^{226}, which has a half-life of 1620 years, and its half-life is 3.825 days. Small amounts of this gas were repeatedly pumped off a radium sample and sealed in small ampoules for radiation therapy. Radon is also a member of the Actinium and Thorium natural radioactive series, and was originally called Actinon and Thoron in those cases. Actinon and Thoron have half-lives of only 3.92 s and 54.5 s. In 1923 it was decided to call all $Z = 86$ atoms Radon.

Radon was the basis of a typical scam not long ago, when a homeowner could pay money to have his house examined for deadly radon, another one of the modern homeopathic horrors. Any house built on granitic rock will show some radon, but if the house is reasonably ventilated, there will not be enough radon to do any damage, except in a very few cases when a house has been built on a radium dump or something similar. Actinon and thoron are even more minuscule hazards. Thoron sparked the expensive and superfluous cleanup of the Weshbach gas mantle factory site in Camden, NJ. This is not to say that breathing any considerable amounts of radon is not hazardous, for it certainly is. It is all a matter of amount. The scam now seems to be cleaning air ducts.

Liquid Helium

If a liquid and its vapour are sealed in a container and heated, the pressure in the container, called the vapour pressure, increases steadily, while the vapour becomes denser and the liquid less dense. At a certain temperature, the distinction between vapour and liquid disappears, along with the free surface dividing one from the other. This is the *critical point*, defined by the critical temperature T_c, the critical pressure p_c and the critical specific volume V_c. Above the critical temperature, no pressure is sufficient to produce a liquid, so that to liquefy a gas, the temperature must be below the critical temperature. The change from liquid to vapour involves the absorption of latent heat, and such a phase change is called *first order*.

For water, whose molecules attract one another, the critical temperature is 647.4K, the critical pressure is 218.3 atm (3210 psi), and the critical density is 0.32 g/cc. By contrast, the critical temperature of helium is 5.2K, the critical pressure is 2.26 atm, and the critical density is 0.0693 g/cc. The critical temperature is so low that helium is very well described as a "permanent gas." Hydrogen has a critical temperature of 33K, critical pressure 12.8 atm, and critical density 0.0310 g/cc, showing how effectively inert its molecules are. Oxygen and nitrogen also have low critical temperatures, 155K and 126K.

One way to produce low temperatures and liquefy gases is by expanding the gas through a porous plug, called *throttling*, from a high pressure to a low one. There is usually a change in temperature, called the *Joule-Kelvin Effect*. Although this is an irreversible process, it is easy to show that the enthalpy of the gas is not changed. The enthalpy H is the sum of the internal energy of the gas, U, and the product of pressure and volume: $H = U + pV$. If we apply the First Law to the process, the heat added, Q, which is zero, is $Q = U - U' + W$, where W is the work done by pistons moving to keep the pressures constant on the two sides, p on the high pressure side, and p' on the low pressure side. Then, we have $0 = U - U' + pV - p'V'$, or $U + pV = U' + p'V'$, or $H = H'$. This makes it easy to predict the change in temperature in the process. An important parameter is the slope of the curves H = constant, or $\mu = (dT/dp)_H$, called the Joule-Kelvin coefficient.

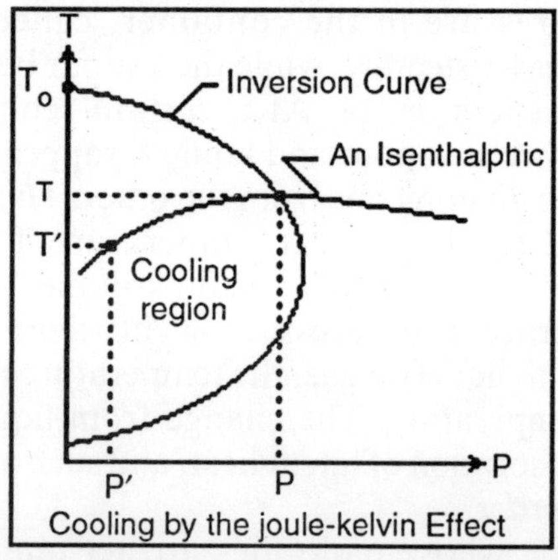

Cooling by the joule-kelvin Effect

Thermodynamics shows us how to express μ in terms of the equation of state of the gas: $\mu = (1/C_p)[T(dv/dT)_p - v]$. If we use the ideal gas equation of state, we find that $\mu = 0$. This means that we cannot cool or heat an ideal gas in this way. For a real gas, μ is positive, meaning that the gas cools on throttling, inside a curve on a T,p plot called the *inversion curve*, as shown in the diagram. It is clear that the gas must be cooled below the temperature T_o before we can cool it in this way. One isenthalpic curve is shown. If the initial state is on the inversion curve, we get the maximum cooling. The state p,T is on the inversion curve, and the final state p',T' is on the H = constant curve through the initial state. There is a cooling from T to T'. To liquefy a gas, we compress it and throttle it, and pass the cooler gas back in a countercurrent heat exchanger to heat the high-pressure gas. The gas becomes continually cooler, and finally it liquefies and can be drawn off.

For air, T_0 is 603K, so we are well within the inversion curve at room temperature. Therefore, it is relatively easy to liquefy air in this way, so this process is the foundation of a considerable industry. Liquid nitrogen separated from the liquid air boils at 77K, and can be used to precool hydrogen below its T_0 of 202K. Then, we are in a region where the hydrogen can be liquefied by the Joule-Kelvin effect. This is more difficult than liquefying air, but can be done, and the liquid hydrogen boils at 20.4K. T_0 for helium is above 25K, so precooling with liquid hydrogen will bring it inside the temperature range where it can be cooled by throttling, in spite of its small nonideality. The result of this rather difficult process is liquid helium.

Another way to cool a gas is to have it do work adiabatically against a piston in an engine, and this has no temperature boundary like the Joule-Kelvin effect. Gas liquefaction using such engines is now common not only for helium, but also for air. These liquefaction plants are small and easily handled, so that every lab can have its own source of liquid air. The Collins helium liquefier, described in Zemansky (see References) makes even liquid helium readily available. Liquid helium is about 500 times more expensive than liquid air. Helium was first liquefied by Kamerlingh Onnes in 1908, using the Joule-Kelvin effect as we have described.

Liquid helium boils at 4.2K, has a density of about 0.125 g/cc, and an index of refraction of 1.024, close to that of a vacuum. The latent heat of vaporization is only 5 cal/g, so it evaporates rapidly and must be carefully insulated to reduce loss. Helium cannot be frozen at atmospheric pressure. Since it is so light, there is considerable zero-point energy, and this unavoidable shuffling keeps it from forming an orderly crystal. Helium can be frozen at higher pressures, however. At 25 atm it freezes into a hexagonal close packed solid near absolute zero. Even hydrogen freezes at 14K. Because helium remains gaseous when all other gases have liquefied, or even frozen, it has been valuable in rocketry, which uses many cryogenic gases as fuels.

A very curious thing happens at around 2.172K. This point is accurately known because the specific heat has a sudden spike at this temperature, resembling a Greek gamma on a plot of specific heat vs. temperature. Therefore, it is called the *gamma point*. Helium above the lambda point is called Helium I, and the phase below the lambda point is called Helium II. Below the lambda point, the atoms begin to associate in a single quantum state with a volume equal to that of the liquid. The fraction of atoms in the superfluid ground state is about $N_s/N = 1 - (T/T_\gamma)^{3/2}$. This curious state is called a Bose-Einstein condensation, and is a single quantum state with a macroscopic extension. The condensation occurs in momentum space, not in configuration space. Each atom in this state is exactly equivalent to any other, and the state is unchanged on the exchange of any two atoms. This many-atom state is possible because the spin of the helium atom is zero, which requires the wave function of multiple atoms to be symmetric on exchange of any two atoms. It is not a simple state, but has many degrees of freedom, and its own peculiar excitations. Liquid helium II can be considered as composed of two interpenetrating fluids, the superfluid and the normal fluid. It will support normal sound waves, but also waves in which the superfluid and normal fluid move in opposite directions, called *second sound*, where there are no pressure fluctuations, but temperature fluctuations, or fluctuations in the ratio of the densities of superfluid and normal fluid, instead.

P. Kapitza found that the superfluid component had zero viscosity in 1938, while Allen and Misener discovered the fountain effect in the same year. Helium II had been studied since 1908, and many of its mysterious properties had been noticed before that time. Landau and Tisza proposed the two-fluid theory of helium II in 1941, which clarified much of what had been observed. Superfluid helium will flow without pressure drop through a fine capillary that will completely stop the normal fluid of independent atoms. Because it attracts itself so little, it will wet any surface and flow over it by capillary action, creating a thin film less than 100 nm thick. Flow in this film can serve as a siphon to equalize levels in different containers. The helium in a suspended cup will even run over the rim and drip off the bottom. If a container of it is rotated, the superfluid will not rotate. However, it may be penetrated by quantized vortices that give the macroscopic impression of a mass rotation; all the angular momentum is in these vortices. The quantum of circulation (line integral of the velocity around a closed path) is h/m, where m is the mass of the helium atom.

Helium II is actually a many-particle system in which some of the atoms are in the superfluid ground state, while the remainder are in excited states, whose energies are functions of their momenta. The excited states are described as phonons, quantized vibrations. When the wavelength of a phonon approaches the average interatomic spacing, a new kind of excitations called rotons appear. (These are not the quantized vortices mentioned above.) The "normal" fluid is represented by the atoms taking part in these excitations. The two-fluid picture is only a convenient expression of these facts, and should not be taken literally.

Liquid helium can be cooled evaporatively by pumping on it. Above the lambda point, it boils vigorously like an ordinary liquid, but when the lambda point is passed is suddenly becomes still and evaporates only from the free surface. The reason for this is the extraordinarily high heat conductivity of the superfluid that quickly equalizes the temperature and does not allow any local variations that produce bubbles of vapour. This is not normal heat conductivity, but is essentially a mass motion of the superfluid. The superfluid has zero entropy (it is a single state, and S = k ln W with W = 1). If helium II is locally warmed, some superfluid must change into normal fluid to take up the entropy dQ/T, since the entropy of the superfluid is zero. This causes superfluid to flow into the warmed region, and normal fluid to flow out. This behaviour gives rise to the striking *fountain effect* of Allen and Misener, and is also related to second sound.

In general, if a warmer container is connected to a cooler one by a capillary, the superfluid will flow against a pressure gradient from the cooler container to the warmer. This flow and conversion transports heat very effectively, and is responsible for the very large thermal conductivity of helium II.

The phenomenon of *superconductivity* was discovered by Onnes in 1911, in mercury cooled to 4.2K. In superconductivity, the electrons form a macroscopic quantum state in which the electrical resistance is precisely zero. In this case, pairs of electrons associate with opposite spins, so that these Cooper pairs have zero spin, and can fall into a Bose-Einstein condensation just as the helium atoms do.

Superconductivity usually requires the very low temperatures that can only be obtained with liquid helium. Helium atoms have no charge, and so are not superconducting, but the zero viscosity is analogous to the zero resisitivity. These matters have led to theoretical investigations that have clarified the structure of matter to a large degree. Superconductivity and superfluidity depend on the reduction of thermal agitation to a very small value.

Until the discovery in 1972 of the Yttrium-Barium-Copper Oxide ceramic superconductors with a superconducting temperature of 90K, all superconductors had to be cooled with liquid helium, a difficult and expensive process. The high-temperature superconductors, with transition temperatures as high as 125K, can be cooled with liquid nitrogen. However, they have not proved a great technological bonanza so far, and commercial use of superconductors for things like electrical transmission is still just a hope. They may be a solution to a not particularly pressing problem.

HELIUM-THREE

Practically all helium is He^4, with two protons and two neutrons very tightly bound. About 1 part per million of atmospheric helium is the stable lighter isotope He^3, with only one neutron. There is only a tenth as much in natural gas helium, confirming its radiogenic origin. No nucleus with just two protons is known, but He^6, which is beta-active with a half life of 0.82 s and changes into stable Li^6, has been found. He^5 has only a very transient existence, almost immediately emitting a neutron.

It is impractical to isolate He^3 from natural sources. Supplies sufficient for laboratory work are obtained from the beta-decay of tritium, H^3, which has a half-life of 12.3 years. If you fill a bottle with tritium, and wait long enough, you end up with a bottle of He^3. One might expect tritium to be the more stable, because of the smaller electrostatic energy, but this is not the case. Tritium turns out to be available because it is used in nuclear weapons, and is manufactured in fission reactors by the capture of thermal neutrons by Li^6, which immediately splits into an alpha particle and a triton, the nucleus of H^3, tritium. It could also be made by neutron capture by H^2, deuterium, but the cross section is 2000 times smaller (0.57 mb vs. 950 mb). The nucleus of He^3 is called a *helion.* A triton, incidentally, combines with a proton at high enough temperatures that the electrostatic repulsion can be overcome to form an alpha particle, with the release of 20 MeV.

The only difference between He^3 and He^4, besides the mass difference, is that the nuclear spin of He^3 is 1/2, making it a fermion, instead of zero. This has the striking result that there is no lambda point as might be expected, and so no superfluid, simply because such a state could not have the necessary symmetry. The mass difference makes the boiling point 3.2K instead of 4.2K, because of the larger zero-point energy for the lighter atom. In 1972, a superfluid He^3 was finally observed, at a temperature below 3 mK. The He^3 atoms associated in pairs with their nuclear spins aligned, to give spin 1, and the resulting bosons could associate into a superfluid.

USES OF HELIUM AND ARGON

The uses of helium depend either on its small atomic mass, or on its chemical inertness. Many applications involve both properties. The uses of argon depend on its inertness. Both helium and argon are available in industrial quantities, and are relatively cheap. The rarer inert gases Ne, Kr and Xe are used in very small amounts in electrical discharges, together with He and A. Xe, as we have remarked, is useful as an anaesthetic.

Helium and argon are used in welding to shield the hot metal from the atmosphere, especially in the case of reactive metals. Classic arc welding uses consumable electrodes coated in a flux. In the heat of the arc, the flux evolves protective gases and forms a liquid slag to protect the metal. In gas metal arc welding (GMAW), helium or argon is blown on the weld, protecting it without producing a slag and excluding oxygen or hydrogen. This seems to have been developed in World War II for welding magnesium, but has now been applied much more generally. It is generally considered necessary for good welds in titanium, as well as for aluminium and stainess steel, where it eliminates weld porosity. In gas tungsten arc welding (GTAW), non-consumable tungsten electrodes are used. Helium is often used as an inert atmosphere in growing semiconductor crystals and for similar processes.

The major use of argon is to fill incandescent lamp bulbs. Although the gas cools the filament somewhat, evaporation of tungsten is discouraged and the life of the lamps is significantly extended.

The small atomic mass leads to large thermal velocities, and this implies rapid diffusion and easy heat transfer. Hydrogen leads in these properties, of course, but helium is not far behind. Helium is used for leak detection in vacuum systems. The gas is blown around likely leak sources, and will diffuse much more rapidly than air through even the smallest opening. The helium is easily detected by an ionization gauge or other means, signalling the presence of a leak. The rapid diffusion makes helium a good carrier gas for gas chromatography. Helium is also used as a driver gas in hypersonic wind tunnels. The high thermal conductivity, and its zero neutron capture cross section, make helium a good coolant in gas-cooled nuclear reactors, though its low density works against it.

Helium gained fame as a lifting gas, and we have seen that this spurred the large-scale production and conservation of the gas, in spite of its inherent rarity. We have already remarked that it is nearly as effective as hydrogen in this respect, as 25 is to 27. The net lift of a balloon is the weight of the air displaced, less the weight of the balloon, lifting gas and ballast. All balloons using hydrogen or helium are strongly affected by the rapid diffusion of these gases, which soon leads to a decrease in lift, which is compensated by throwing ballast overboard. Incidentally, the lifting gas is contained in separate bags like the compartments of a ship, so that the compromise of one is not the loss of all. The use of helium in small balloons instead of hydrogen seems to be a frivolous waste of a valuable substance. Of course, the reason it is used is the elimination of the hazards of hydrogen combustion, perhaps more in the gas sources than in the actual balloon. If the predictions of a "hydrogen economy" do indeed come true, we will hear more about the dangers of hydrogen. All the problems of working with helium arise with hydrogen as well, with the added excitement of explosion.

The inert gases are excellent as filling gases for electrical discharges, since they do not react with the electrodes, and tubes containing them have long service lives. Neon and argon are used in glow discharge lamps, where the light comes principally from the negative glow near the cathode. Neon has its typical orange-red colour, while argon produces a violet. The electrodes are specially treated to encourage electron emission, so that the striking voltage of the lamps is low. "Neon lights" use the positive column in a long tube as the light source. Helium gives a yellow light, and neon red-orange. Mercury with argon to aid starting gives blue. Colored glass tubing is used to create other colors. The pressure is a few mm Hg, and the voltage drop around 100 V per metre. These lamps are supplied with a high-voltage transformer with large leakage reactance so that the voltage is high for striking, and decreases when current flows. The current may be 25 to 50 mA.

A discharge through a mixture of helium and neon creates a population inversion in the neon that can be used in a laser. The helium is ionized in the discharge, and when the electrons recombine and cascade down in energy, they pile up in the two metastable states $2s^3S$ (19.82 eV) and $2s^1$ (20.62 eV) of the electron configuration 1s2s. Fast transitions to the ground state are discouraged by the selection rules, in one case forbidding triplet-singlet transitions and in the other S-S transitions. By good luck, the energies of each of these metastable states is about equal to the energies of two upper levels of laser transitions in neon. The transfer of energy is quite likely, populating these states. One state makes a transition of wavelength 632.8 nm, the familiar He-Ne visible red laser line, to a lower state, while the other makes a transition of 1152.3 nm to the same lower state. This lower state is kept empty by allowed transitions of wavelength 594.5 nm to a still lower state that is depopulated by collisions with the walls of the discharge tube.

All of the inert gases have a large energy gap between the ground state and the first excited state, corresponding to a transition in the far ultraviolet. The visible lines in the spectra are due to transitions between higher states. This explains why the inert gases are transparent, since ordinary visible frequencies are too low to excite them when they are in the ground state. The helium spectrum in a discharge tube shows strong lines at 447.1 nm (blue), 501.6 nm (green), 587.56 nm (yellow) and 667.8 nm (red). This spread of lines creates a whitish light. Neon has a cluster of lines in the red and orange, while argon has groups of lines in the blue and in the red.

Chapter 4

Uranium

Uranium is now indelibly associated with atom bombs and nuclear power, but was a chemical curiosity of little importance for a hundred years after its discovery. Then, it led to the discovery of radioactivity and the elucidation of the structure of the atom. A half-century more saw its use in nuclear explosives and in power generation. Most of the importance of uranium has been in physics rather than in chemistry, unlike that of most chemical elements. In this article, as much as I can find on the chemistry of uranium will be outlined, but the physical subjects of radioactivity and nuclear power will be discussed at length. The elements of these subjects are not hard to understand, and can go a long ways towards a useful appreciation of subjects that are often in the news. In particular, it may help in making a meaningful and intelligent assessment of the risks and advantages in these matters.

Martin Klaproth, an early convert to Lavoisier's chemistry, discovered a new substance in pitchblende from Joachimsthal, Bohemia in 1789, and named it *uranium* after the ultimate planet in the solar system as then known, Uranus, which had been discovered by William Herschel a few years earlier, in 1781. Since Mendeleff's periodic system did not yet exist, Klaproth would not have realized that uranium, atomic number 92, would for many years be the ultimate element. When elements 93 and 94 were made many years later, the names *neptunium* and *plutonium* carried on the analogy with the solar system. Klaproth had actually found the oxide, not the metal, but at least he proved that pitchblende was not an ore of tungsten, iron or zinc. E. M. Pıligot isolated the pure metal in 1842. Uranium was found to have the atomic weight of 238.07, the greatest of any element, and was assigned the symbol U.

A fresh surface of metallic uranium is white and lustrous, with a bluish tinge, but it soon tarnishes in air. There are three crystalline forms of metallic uranium; the usual form is orthorhombic, and is malleable and ductile. It can be cast, rolled and extruded, and is often produced in 1"-diameter rods. The metal oxidizes rapidly when heated to 200°C, and burns in air. This behaviour is typical of all the heavy metals in this region of the periodic table, including radium and plutonium. When used as metals, they must be "cladded" with another metal, such as zirconium or aluminium, to protect them from moisture and oxygen. Uranium is soluble in mineral acids, and will displace hydrogen from water. Since the nitrates of uranium and its friends are soluble, chemistry is generally performed with them. A notable property of uranium is its high density, about 19 g/cc (sources give 18.9 and 19.04). This is about twice the density of lead (11 g/cc), and puts uranium up with the densest metals, such as platinum, osmium and iridium. Uranium melts at 1133°C with latent heat 2.5-3.0 kcal/mol, and boils at 3925°C with latent heat 110 kcal/mol. Its coefficient of thermal expansion is 15×10^{-6} per °C, and its electrical resistivity is 29 μΩ-cm.

The availability of metallic uranium as a by-product of the production of enriched uranium, and of plutonium for nuclear weapons ("spent uranium") has encouraged its use in armor-piercing projectiles, and perhaps as armor itself, taking advantage of its high density. The uranium in these devices escapes its cladding when the projectile melts on impact, releasing burning uranium which can do no one much good.

Uranium has a ground-state electron configuration of $5f^36d7s^2$ outside a radon core. The f, d and s levels all have about the same energy, so it is not easy to guess what orbitals are used. Three of the valence electrons occupy the f orbitals that were left empty, moving from the d orbitals where they might be expected to reside. The first ionization potential of the atom is about 4 V, a relatively low value. Uranium is, therefore, an electron donor, easily oxidized, and exhibits valences of +3, +4, +5 and +6, with +4 (uranous) and +6 (uranyl) being the most prominent. Uranium is more liberal with valence states than most of its neighbors. The oxides are not soluble in water, so we do not know if they would give basic or acidic hydroxides, but basic would be more probable, so that uranium would definitely behave as a metal. The uranium atom is not unusually large, since the high nuclear charge sucks the electrons in closely. The ions are only about 0.1 nm in radius, a very typical value. Natural uranium is, on the average, an isotopic mixture of 99.27% U^{238}, 0.72% U^{235} and 0.0055% U^{233}.

UO_2, UO_3 and U_3O_8 are observed as oxides. Uranium exercises all its valences in combining with the halogens. The compound UF_6, uranium hexafluoride, is typical. UF_6 is a means of "gasifying" uranium, since it boils at 56.2°C at 65mm pressure, and is used in isotope-separation processes such as diffusion and centrifugation. Uranium does not seem to combine with hydrogen, but it forms the sulphides US_2 and U_2S_3, which ignite rather readily. Thre is also a uranous sulphate, $U(SO_4)_2 \cdot 4H_2O$, which is nearly insoluble but dissolves in HCl. Uranous compounds are usually greenish or dark.

When the oxides are dissolved in acids, the stable uranyl cation, UO_2^{++}, is formed. This cation forms numerous soluble salts, such as $UO_2(NO_2)_2 \cdot 6H_2O$. The yellow, orthorhombic crystals melt at 60.2°C, boil at 118°C and have a density of 2.807 g/cc. Deliquescent uranyl chloride, UO_2Cl_2, and uranyl iodate, $UO_2(IO_3)_2$, are also known. Uranyl gives the colour yellow in aqueous solution, and this colour is typical of all hexavalent uranium.

Before the discovery of radioactivity, uranium was a curiosity with few uses. It gives a green colour to ceramics and glass, as shown by the borax bead test. Uranyl nitrate is used in a reddish-brown photographic toner. Zinc uranyl acetate ("sodium reagent") is used to precipitate sodium in qualitative analysis, since pale yellow $NaZn(UO_2)_3(C_2H_3O_2)_9$

is one of the very few insoluble sodium compounds. Uranium metal was used for some gas-discharge tube electrodes, and compounds were used in making incandescent gas mantles. None of these uses is current, and it seems to be difficult even to procure uranyl compounds from chemical supply houses, probably more from lack of use than uranophobia. Uranium makes hard carbides like tungsten and molybdenum, which it resembles chemically. However, its use in hgh-speed tool alloys gave no advantage.

Uranium and its compounds are poisonous, and should not be taken internally, nor should their dusts be breathed. The chronic breathing of dusts is probably the most hazardous activity, but unless you work in the uranium-mining industry without a mask, you will not be exposed to this danger. Natural uranium minerals do not represent a hazard, since they are very insoluble and only very weakly radioactive. They can be collected, handled and studied with complete safety. Even uranium metal, natural or enriched, can be handled safely like any other metal. Granites usually contain a very small amount of widely-disseminated uranium, which generates the heavy inert gas radon. If this gas is trapped, in rare cases it may be present in such concentrations that breathing it is a hazard. The radon will decay in the lungs to a chain of radioactive substances that can do damage. This is an extremely feeble hazard that is completely avoidable by proper ventilation. The hazards of radioactivity will be discussed in more detail below.

Mineralogy of Uranium

The abundance of uranium in the earth's crust is about 4×10^{-4}%, roughly the same as that of boron or tungsten. It is a little more common than gold or rubidium, but thorium and lead are 3 and 4 times more abundant, respectively. Boron and tungsten are found in common minerals, boron in Kernite and tungsten in Ferberite-Heubnerite, but no uranium mineral is common, and most are very rare. The availability of an element depends not only on its abundance, but on its concentration by geological processes into an economic ore.

The demand for uranium as a weapons material led to extremely thorough prospecting, and its occurrence is now well-known. There are probably only some 2,000,000 tons of high-grade ore available, a very small amount for a material of wide use.

The most common uranium ore is *pitchblende*, which contains mainly uranous oxide, UO_2, which as a pure mineral is *uraninite*. The "pitch-" came from its black, resinous appearance and botyroidal habit, which looks like pitch, and "-blende" because it cannot be smelted to any valuable industrial metal, like copper. It is a variable material, since the oxide is more or less partially oxidized to U_2O_8, which decreases the hardness from Mohs 5.5 down to 3-4, and its density from a very high 9.0-9.7 down to 6.5-8.5, which is still unusually heavy for a mineral. It is soluble in nitric, sulphuric and hydrofluoric acids, but not in hydrochloric.

Pitchblende was best-known from Joachimsthal, Bohemia in the Erzgebirge (now Jachymov)—this was the source of uranium for Klaproth, and later for the Curies. It is found also in Germany, Portugal, and notably in the Congo, at Lumumbashi in Katanga.

It was known in Cornwall, found in Colorado at Central City, in Canada around the Great Bear Lake in the northwest, while large crystals come from Wilberforce, Ontario. Because pitchblende contains uranium, it also contains all the radioactive decay products of uranium, down to lead. The intermediate products are all present in constant proportions, while as uranium decays the lead builds up. This is true of any uranium mineral, not just pitchblende, of course. Because the half-life of uranium is so great, its daughters are present in very small concentrations.

The other economic ore of uranium is *carnotite*, with an average composition of $K_2O.2U_2O_3.\ V_2O_5.2H_2O$, a soft, yellow, powdery mineral that is much easier to work with than pitchblende, but rather rarer. A specimen is shown at the right. It's a powdery coating with the typical colour. Its hardness is 2-4, its density 4.136 g/cc. Its monoclinic (or orthorhombic?) crystals are biaxial, with rather high indices 1.75 and 2.06. Normally, however, it is not found as euhedral crystals, but as crusts and cementing sandy sediments. Its solution in HCl is yellow, in HNO_3 green, and in H_2SO_4 orange. It is a secondary mineral concentrating widely-disseminated uranium in granites. It is found in Utah and Colorado, notably in the Paradox Valley, and in Wyoming near Jeffrey City.

The mineral Coffinite, probably related to Carnotite, was discovered in 1955 in Colorado, but does not appear in common mineralogical references. The unusual violet *Ianthinite* is $UO_2(OH)_2$. Curite is said to be $3PbO.8UO_3.4H_2O$. There is a number of very rare uranyl minerals, that are copper-uranyl or calcium-uranyl phosphates, vanadates or arsenates. Among the silicates are Kasolite, $PbUO_2SiO_4.H_2O$, Uranophane, $CaH_2(SiO_4)_2UO_2.5H_2O$, and Cuprosklodowskite, $Cu(UO_2)_2Si_2O_6(OH)_2.5H_2O$. The structures of these rare minerals are not clear to me, so I simply report them. They do not appear in Dana, and probably do not represent practical sources of uranium. They do make very attractive colored crystals, however.

Commercial production of radium began in 1910. About 750 tons of raw pitchblende, or 12 tons of concentrated ore, must be processed to win one gram of radium. From 1913 to 1925 the chief source of uranium was Colorado carnotite. In 1925, discovery of pitchblende in the Congo diverted production to that source. In 1931, Canadian pitchblende from Great Bear Lake took over. The uranium boom of the 1950's saw uranium mills in Moab, Utah; Jeffrey City, Wyoming; Durango, Canon City and Uravan, Colorado; and many other places in the arid West, which processed mainly low-grade carnotite into oxide. These mills are now derelict, or scratch out an existence as best they can, beside huge piles of waste.

These mines and mills produced only weakly radioactive concentrates ("yellowcake"), not metallic uranium or purified radionuclides.

Vanadium and uranium were concentrated at Durango, Colorado from 1942 to 1962, resulting in two large tailings piles at the southwestern edge of town. There are 23 other such sites in the West. These piles were unsightly, and leachings from them might be chemically obnoxious, but the radioactive hazard was extremely small. Any house built on them might be suspected for radon buildup, and the dust should not be chronically breathed (like so much nonradioactive dust), but they contained much less uranium than the original carnotite did, and presented even less of a danger. Indeed, if only vanadium had been extracted, which was actually the primary purpose, they would have attracted no interest, except by those who had to look at them. Uranophobia and ignorance, however, have led to expensive follies. At least they could get the government to pay for taking them out of sight. The heaps, I must assume, have now been moved to somewhere where they are also out of mind. Transportation accidents while moving the piles probably killed and injured more people than ever the dirt will do.

Thorium is usually found associated with the rare earths, especially Cerium. The principal source is Monazite sand, a Th-Ce-Y-La-Nd phosphate, which can contain up to 12% ThO_2. Monazite is heavy and hard, so is easily separated from the lighter silicic gangue. A very rare mineral is Thorianite, $(Th\text{-}U)O_2$, which contains uranium as well. Brannerite and Betafite are other rare minerals containing rare earths, thorium and uranium, but are not common enough to be ores.

RADIOACTIVITY

Henri Becquerel (1852-1908) was studying the properties of fluorescent compounds in 1896, one year after Rφntgen's amazing discovery of X-rays. He noticed that a uranyl compound, obtained from carnotite, fogged photographic emulsion that was wrapped to exclude light. He was, of course, delighted to have discovered a "radiation" himself, and set

Marie S. Curie (1867-1934) to work on it. She was aided by Pierre Curie, her husband and a noted physicist, who quickly became interested as well. They observed that when the uranium was separated from pitchblende, the residue was very active. Curie named the phenomenon *radioactivity*. At length, the radiation was found to consist of positively charged heavy particles, negatively charged electrons (themselves only recently discovered), and electromagnetic radiation like X-rays but more penetrating. The three components were called α, β and γ radiations, respectively. Becquerel and the Curies received the Nobel Prize in Physics for 1903. Marie Curie received the Chemistry prize in 1911. There was even a Hollywood film of the 1940's about her discovery of radium.

Curie discovered that the pitchblende evolved small amounts of the gas helium, identified by its spectrum, which had been found in the sun some years earlier. This was the first terrestrial helium found. All helium in the atmosphere and in natural gas comes from radioactive decay, since it was soon demonstrated that the alpha particles became helium. This led Rutherford and Soddy to deduce that radioactivity was the result of the transformation of one atom into another in 1902, and from this to the concept of the nuclear atom. Radioactivity was the result of nuclear transformations. In 1913, Rutherford demonstrated the existence of nuclei, and Bohr showed how electrons moved. Alpha particles were simply the nuclei of helium atoms.

A nucleus can be characterized by its atomic number Z, the positive charge on its nucleus in units of the electronic charge, and its mass number A, the total number of *nucleons* in its nucleus. The nucleons are the *proton*, p, with mass 1.007596 amu and charge +1, and the *neutron*, with mass 1.008986 amu and charge 0. The atomic mass unit, amu, is just the reciprocal of Avogadro's number, and corresponds to 931 MeV in energy, according to Einstein's relation $E = mc^2$. An MeV (mega-electron-volt) is 1.6×10^{-13} J. A nucleus (Z,A) contains Z protons and $N = A - Z$ neutrons. Nuclei with the same Z, and therefore the same chemical properties, are called *isotopes*. The isotopes of uranium cover a range in mass number from 227 to 240. U^{227} is a 6.8-MeV alpha-emitter with T = 1.3 min, and U^{240} is a beta emitter with T = 14 hr.

The free neutron is heavier than a free proton, which favors *beta decay* of the neutron, $n^0 \rightarrow p^+ + e^- + \nu^0$. The ν is an *antineutrino*, necessary so that the fermion number (electron +1, antineutrino -1) is conserved (zero on both sides of the equation). The antineutrino carries off varying amounts of energy, so the maximum energy the electron can have corresponds to a zero antineutrino energy. Since the mass of the electron is 0.000548 amu (0.51 MeV), an amount of mass of 0.000842 amu has disappeared in this reaction, which must appear as the kinetic energy of the products. Most of this energy will be in the electron, since the proton is much heavier, and the amount will be 0.78 MeV.

The chance that a neutron will decay in any time interval dt is proportional to dt. If N is the number of neutrons, this means that $dN = -\gamma N dt$, since the number decaying in dt will be proportional to the number present. The constant γ is called the *disintegration constant*, and is independent of all physical influences. This is a remarkable law, by no means trivial or obvious, which says a lot about nuclear processes. This equation can be integrated to the familiar $N = N_0 e^{-\gamma t}$, which expresses the exponential decay of a radioactive nucleus. Half the original number of nuclei remain after a time $T = \ln 2/\gamma$, which is called the *half-life*. The same answer is obtained no matter when we begin looking. The neutron has a half-life of about 13 minutes, so $\gamma = 0.00128$ per sec. If we have a thousand neutrons, then in the next second about one will transform into a proton. We speak of "disintegration" and "decay" but these are mere vivid terms—nothing disintegrates or decays in any meaningful sense. Half-lives vary from longer than can be measured, billions of billions of seconds, to less than can be measured, to mere nanoseconds. They are absolutely constant.

In the active residue from pitchblende, Curie isolated polonium in July later 19th century, and radium in September. Polonium (Z = 84) is chemically similar to bismuth, radium (Z = 88) to barium. All nuclei with atomic number 84 and greater are unstable, the ones up to about $Z = 100$ are radioactive, the remainder fission spontaneously. Only nuclei up to $Z = 92$ are found on earth, because any beyond this atomic number, if originally present, would have decayed by now to nuclei with $Z < 84$. Those with Z between 84 and 92 are found naturally only because they result from the decay of long-lived nuclei of uranium and thorium. U^{238} (4.67×10^9 y), U^{235} (7.13×10^8 y) and Th^{232} (1.39×10^{10} y) are the three patriarchs that give rise to three sequences of alpha and beta decay ending in Pb^{206}, Pb^{207} and Pb^{208}, respectively. The abundances of these isotopes of lead are 26%, 21% and 52%. Not all lead is of radioactive origin, but much is, and the atomic weight of lead varies depending on the source. The isotope Pb^{204}, not the end of any radioactive series, has an abundance of only 1.3%.

The members of a radioactive series can be detected much more readily by radiation measurements than by chemical analysis, so the chemical nature of many of the daughter nuclei were not originally known. They were given designations like UZ (Pa^{234}), RaF (Po^{210}) or ThX (Ra^{224}). The series often took alternative routes, with α,β or β,α alternatives. In equilibrium, every member of a radioactive series decays at the same rate it is created. Therefore, if N_1 and N_2 are the equilibrium number of two species with disintegration constants γ_1 and γ_2, $N_1\gamma_1 = N_2\gamma_2$, or the numbers are in inverse ratio to the disintegration constants. Those members with the longest half-lifes will be present in the greatest amounts. See if you can work out the ratio of the amount of radium to the amount of U^{238} in pitchblende. Answer: 0.34 g to 1 metric ton of uranium.

The ratio of uranium to lead in a rock can be used to measure the time since the rock was formed, if we assume all the lead came from radioactivity. If r is the ratio of the number of atoms of U^{238} to the number of atoms of Pb^{206} at time t, then the law of exponential decay can be used to find $t = T [\ln(1 + 1/r)/\ln 2]$, if T is the half-life, 4.5×10^9 y for U^{238}. If a particular rock gives $r = 2$, then $t = 2.6 \times 10^9$ years. This method was suggested by Boltwood in 1905, soon after the discovery of radioactivity, and was brought to perfection by

Arthur Holmes in the 1950's. Similar dating methods using the helium or argon trapped in igneous rocks can also be used. Thanks to uranium, we now know the absolute geological time scale, and can extend it back to Precambrian rocks. The oldest rocks found are a bit older than 3,000 million years old, while the age of the earth's crust is about 4,500 million years.

A fourth series would descend from U^{233}, but its half-life is only 1.62×10^5 y, so it all would have been gone long ago. The end of its series is Bi^{209}, whose isotopic abundance is 100%. However, this is not really the end, since Bi^{209} is estimated to have a half-life of some 3×10^{17} y against alpha decay to Tl^{205}, which is stable and would be the end. Not nearly enough time has elapsed for this to have taken place, and there is still lots of bismuth. All isotopes of bismuth appear to be radioactive.

There are a few natural radioactive nuclides not included among the heavy elements and their radioactive series. The rare isotope K^{40}, with abundance 0.0119%, is beta and gamma active with half-life 1.3×10^9 y. The energy of the gamma rays is 1.46 MeV. Rb^{87}, abundance 27.8%, is also beta active with half-life 4.3×10^{10} y. These elements are valuable geologic clocks. Nd^{144} and Sm^{157} are alpha emitters, while Lu^{176} is a beta and gamma emitter. The radioactive nuclide C^{14} is produced by the $N^{14}(n,p)C^{14}$ reaction in the atmosphere due to cosmic rays, and is present in all living matter. Its half-life is 5600 y, and it is a beta emitter. One part in 10^{12} of atmospheric carbon is C^{14}, and results in an activity of 15.3 dis/min per gram of carbon. After death, the carbon is no longer replaced by metabolism, and the activity decreases with the half-life of C^{14}. In 1951, Willard Libby discovered how to use this effect for the dating of carbon-containing fossil matter. One of the first results was the dating of the last continental glaciation to 11,400 years before the present from the trees that were knocked down by the advancing ice. Comparison with other dating methods has shown that cosmic radiation intensity has been constant for at least 30,000 years. However, human activities have upset the C^{14} concentration. The carbon from fossil fuels has no C^{14}, while C^{14} has been produced by nuclear weapons tests. C^{14} is present in such small amounts that these additions produce a notable effect, and must be allowed for, especially when current material is used as a standard. Radiocarbon dating can be quite accurate in determining the age of artifacts.

Ra^{226} has a half-life of 1620 y, corresponding to a disintegration constant of 1.354×10^{-11} per s. Alpha particles of 4.78 and 4.59 MeV are emitted (one or the other), and gamma rays of 0.187 MeV energy. It is the longest-lived member of its radioactive series, except for two uranium isotopes. One gram of radium contains $6.02 \times 10^{23}/226$ nuclei, so the number decaying in a second will be 3.7×10^{10}. This number is called a *curie*, a measurement of the number of disintegrations per second in any radioactive sample. An actual sample of radium that has been around a while is in equilibrium with all of its daughters, each of which contributes its curie. Radium was used in luminous watch dials in the ratio of one part to several thousand parts of the scintillator hexagonal ZnS, that emits a flash of light for every alpha particle that hits it. Zinc blende gives 7 times more light than the next most sensitive scintillator, Barium Platinocyanide. The method of making luminous paint is described in Glazebrook (see References). This was not a dangerous use of radium, because of the extremely small amounts used, but it is probably discouraged at present. The hazard to the user is totally negligible; that to the manufacturer can be avoided with care. Electroluminescent panels are a good substitute with no radioactivity hazard at all.

The result of the decay of Ra^{226} is Rn^{222}, with half-life 3.825 days. If a gram of this gas were separated from the radium, it would have an initial activity of one curie. In radium therapy, this is what is done, and a little radon in a capsule is used instead of the actual radium, and its activity increases with time. The gamma rays emitted by the daughters of the radon provide the therapeutic effect, damaging cancer cells more than healthy cells. None of the alpha or beta radiation can penetrate the metal capsule.

Radioactivity is just a search for stability. For any Z, there is an optimum value of N that gives greatest stability. For small Z, Z = N is preferred, but as Z increases, N > Z is preferred. A nucleus that has too many neutrons can increase the Z/N ratio by emitting an electron in beta decay. This electron does not exist in the nucleus, but is created along with an antineutrino at the moment of decay. A nucleus with too few neutrons can get more by emitting a positive electron, or *positron*, along with a neutrino. The positron is the *antiparticle* of the electron, with fermion number -1, while the neutrino has fermion number +1. Nuclei of intermediate values of A, about A = 55 (in the vicinity of Fe), are more stable than those with smaller or larger mass. Heavy nuclei can move down the line of stability by emitting an alpha particle, reducing Z and N each by two. Since the line of stability curves upwards, this leaves the nucleus with too many neutrons, so generally a negative beta decay follows to set this right. The radioactive series are simply chains of these decays looking for a more stable state. Lighter nuclei, of course, have no alternative like inverse alpha decay that would make them heavier, and beta decays would get them nowhere.

NUCLEAR POWER

Consider building up an atom of U^{235} by assembling 92 protons, 92 electrons and 143 neutrons. The total weight of these constituents is 237.0343 amu. If an atom of U^{235} is weighed, it turns out to weigh 235.1175 amu. What happened was that the parts attracted one another, and we were able to extract a large amount of energy in assembling them. From the mass defect, this energy turns out to be 1785 MeV, or and average of 7.59 MeV for each nucleon. Atoms with A about 55 show a maximum of 8.7 MeV per nucleon. Could we somehow split up the nucleus optimally into smaller parts, we would have $(8.7 - 7.6) \times 235 = 259$ MeV to spare. If the nucleus just splits into two roughly equal parts, we get about 200 MeV. 1 MeV per atom corresponds to 2.3×10^7 kcal/mol, so 200 MeV gives us 4.6×10^9 kcal/mol, or about 2×10^7 kcal/g, enormously more than any chemical reaction. The heat of combustion of petroleum is about 10 kcal/g, for comparison.

This process is called *fission*. Fission was first verified by Hahn and Strassman on 6 January 1939, though Fermi had been working on it since 1934.

At the other end of the mass scale, the mass of a helium atom is 4.003860 amu, while the mass of two hydrogen atoms and two neutrons is 4.035087 amu. This means that we will liberate about 29 MeV for each helium atom we make from protons and neutrons. The binding energy in the helium nucleus is about 7.27 MeV/nucleon, not much less than the maximum possible, so little is gained by further assembly. The energy output is 6.71×10^8 kcal/mol, or 1.7×10^8 kcal/g, even more than we obtained from fission. This process is called *fusion*. The stars get their energy from fusion of hydrogen into helium, so solar energy has its source in fusion. The mechanism involves several intermediate steps, so it is not quite as simple as it might appear. Fusion has been produced on earth by using a fission explosion to ignite a blanket of lithium deuteride, LiD, to fuse two deuterons, each consisting of a proton and a neutron, into a helium nucleus, in what is called a "hydrogen bomb." Attempts to make a fusion reactor began with Project Sherwood in 1948, and success has not yet been achieved. It is certainly possible, but we do not know how to do it.

The heaviest nuclei are on the verge of fission. They wobble like a large drop of mercury, and if given a little extra energy they can be pushed over the edge and separate into two parts. The extra energy can be obtained by the absorption of a neutron, which can approach and enter the nucleus easily because it is not charged, and will not be repelled as a proton would be. A few nuclei will fission if they absorb a neutron of any energy; these nuclei are called *fissile*. The only natural fissile nucleus is U^{235}. When caused to fission by absorbing a slow neutron, it breaks into a light fragment with mass number 80-110, and a heavy fragment with mass number 125-155. The most probable masses are 95 and 139, respectively, and the full range is 72-160. These *fission fragments* possess about 80% of the 200 MeV released as kinetic energy, which they quickly distribute as heat. About 10% of the energy appears as gamma radiation, kinetic energy of emitted neutrons, and β-ray energy. The final 10% is carried away by neutrinos to an infinite distance.

The fission fragments are excessively neutron-rich for their masses, and release several neutrons immediately, on the average, and a few more at a slightly later time. These neutrons are very important, because they can be used to cause further fissions. For a sustained chain reaction, only one neutron need survive to cause a fission from all those produced. Neutrons may leave the mass of uranium, or be absorbed in some non-fission process. The neutrons are emitted at high speeds as *fast neutrons*. They lose energy by elastic collisions with light nuclei, such as hydrogen, deuterium, lithium, or carbon, and are slowed down to become slow, or *thermal* neutrons if they escape absorption. Thermal neutrons are much more likely to cause fission in a fissile nucleus provided they are absorbed in the fissile nucleus. They are also much more likely to be absorbed without profit. The material used to slow down the neutrons is called the *moderator*. It must have a small nuclear mass, and a small cross-section (probability) for neutron absorption, which are very restrictive conditions. The best practical moderators are heavy water (deuterium oxide) and graphite (carbon). The absorption of fast neutrons is rather improbable, except for a critical velocity range for "resonance absorption" that most neutrons escape if they pass through it rapidly.

An assembly of uranium and moderator is said to be *critical* if just one neutron remains to cause fission from those released in any fission. Natural uranium is 99.3% U^{238} and only 0.7% U^{235}, so it is a very delicate matter to achieve criticality. It can only be done with the uranium fuel in lumps, with a moderator of heavy water or graphite. A reactor assembly at Berlin using heavy water failed to achieve criticality in 1939. A group headed by Enrico Fermi succeeded in Chicago on 2 December 1942. This reactor consisted of 385 tons of graphite blocks, with 40 tons of natural uranium oxide in spherical lumps, carefully piled up. Cadmium has a very high cross section for neutron absorption, so cadmium rods were used for control. One rod could be released if an ionization chamber detected large amounts of radiation, another was tied with a cord that could be cut if necessary by an assistant on a balcony, and a third was intended for control by being moved in and out to change the reactivity. The "pile" went critical without incident, and operated at a power level of 2 kW, cooled by the natural convection of air. Control of the reactor was facilitated by the existence of the "delayed" neutrons that were essential for criticality, and by the time required to moderate the neutrons to thermal speeds.

It is impossible to construct a nuclear explosive with natural uranium, since a nuclear explosive must be "fast critical" and this requires more neutrons than can be produced by 0.7% of fissile material. The uranium must be "enriched" by increasing the proportion of U^{235}, a very difficult matter. The method used was gaseous diffusion, relying on the more rapid diffusion of the lighter U^{235} relative to U^{238}. The difference is tiny, so the most heroic measures are necessary, involving thousands of diffusion stages to get an even modest enrichment. Nevertheless, large quantities of enriched uranium were produced in case it was the only way to get fissile material.

One of the things that can happen to a neutron slowing down in a reactor is that it can be absorbed by U^{238}. If the neutron is a fast one, it may cause fission in U^{238}. Nuclei that will fission when they absorb a fast neutron are called *fissionable*. Not only U^{238} is fissionable, but also Th^{232}, as determined by Nishina in 1939. Fast fission contributes to the reactivity of a thermal reactor, but only by a small amount. It is more likely that what is now U^{239} after eating the neutron will become Np^{239} by beta decay (T = 23.5 min), which then becomes Pu^{239} by a second beta decay (T = 2.33 days). Pu^{239} has a half-life of 24,300 years, so it will stay around for a while. By a stroke of luck, it happens that Pu^{239} is fissile, like U^{235}.

When a reactor runs, fission fragments build up in the fuel elements. Some of these fragments have large neutron absorption cross-sections, and so are called *reactor poisons*, since they lower the reactivity. Very soon, the reactor would become subcritical and the fires would go out, unless the fuel elements were withdrawn and replaced by new ones. When this happens, only a few percent at most of the fuel has been consumed. This "spent fuel" is not by any means "nuclear waste," as so commonly described. The fission fragments must be cleaned out, and the fuel made into new fuel elements for further burnup. The first thing is to let the "hot" fuel elements cool down for 100 days or so, usually under water, until they can be handled. They are, however, still extremely radioactive and must be carefully managed.

The fuel elements are chopped up and dissolved in nitric acid. Then the solution is subjected to a countercurrent organic solvent extraction with n-tributyl phosphate diluted with kerosene. This solvent absorbs the +6 uranium and +4 plutonium components, leaving all the rest, fission fragments and all, in the nitric acid solution. The +4 plutonium is then reduced to +3 with ferrous sulphamate (the sulphamate removes the nitrite that was used to make sure the plutonium was in the +4 oxidation state) and back-extracted with water. This recovers all the plutonium as an aqueous solution. The uranium is left in the solvent and extracted with dilute nitric acid into an aqueous solution. The solvent is purified and recycled. Now we have three vats. One contains the highly radioactive fission fragments, another the uranium, and the third the plutonium. These two vats are hardly radioactive at all. The fission fragments are disposed of (easier said than done), the uranium precipitated and made into new fuel elements when combined with a suitable amount of fissile material, which can be some of the Pu^{239} also recovered. A thermal reactor burning U^{235} can often make as much Pu^{239} as it burns, becoming a *pseudo-breeder*. However, a fast reactor is usually necessary if an excess of Pu^{239} is required. A reactor that burns Pu^{239} and makes more than it burns is called a *true breeder*. The U^{238} that is turned into Pu^{239} is called a *fertile* species. Only one process, the "Purex" process, has been described here, but others are similar and have the same end results.

After the CP-1 reactor proved the principles valid, it was

decided to build large, water-cooled graphite-moderated reactors at Hanford, Washington to be used as pseudo-breeders to make plutonium. This was much, much easier than isotope separation to obtain fissile material, since a far smaller degree of enrichment could be used in the fuel, and the product was, in effect, 100% enriched. A test reactor was built at Oak Ridge, Tennessee that went critical in November 1943. This reactor was forced-air-cooled, operating initially at 1 MW, later at 4 MW, and used to test the materials for the Hanford reactors, which went into operation in September 1944.

There is also insufficient space here to give a proper account of the use of reactors for power production. The reactor serves only to generate the heat to make steam (or boil mercury); once this is done, the power plant is exactly like one burning coal. Much of the world concentrated on reactors using natural uranium, like the water-cooled, heavy-water moderated Canadian reactors, or the CO_2-cooled, graphite moderated British reactors. The United States was crawling with enriched uranium, so American reactors could use light-water moderators and other luxuries, with plenty of excess reactivity.

The first reactor to enter commercial service was the pressurized-water, highly-enriched uranium reactor at Shippingport, PA, on 2 December 1957. It resembled the smaller reactors used in nuclear submarines that were developed at Arco, Idaho. From 1954 until 1971, enriched uranium was leased to power plants, apparently so that it could be returned to extract the plutonium for weapons. Without this considerable subsidy, nuclear power plants using enriched uranium would be economically impossible. By 1971, the need to enlarge the huge plutonium stockpile seemed unnecessary, and public resistance to nuclear power had rendered its future dim. Many efforts, like the Advanced Gas Cooled Reactor at Ft. St. Vrain, Colorado, were utter failures because of engineering deficiencies, though the utilities originally envisioned money flowing like water from the government. Since 1971, uranium has been available for purchase by utilities. It is not sold at anything near the cost of its production. One very significant change is that the processing of spent fuel has ceased in the United States, and is not encouraged anywhere in the world. Instead of the horribly radioactive waste tip of reprocessing, we now have the horribly radioactive pile of spent fuel under water, which, as we have pointed out, is certainly not waste, just fuel of which 2% has so far been used. The danger, more than that of the radioactive waste, seems to be the danger of using the plutonium that results from reprocessing for unauthorized nuclear explosives.

Without reprocessing, the limited reserves of fissile material will soon be used up, and then nuclear power is gone for good. Reactors themselves are safe, clean and nonpolluting. They compete mainly with coal as a fuel, and several centuries of cheap coal are still available, even at the current rates of consumption. They do nothing to reduce dependence on cheap petroleum, however, and that will be exausted much sooner. The greatest problem is the huge amount of radioactive fission products that will accumulate, and proper management of this resource seems well beyond the capacity of modern politics and politicians. Even the casual reader of this article now knows more about nuclear power than the whole lot of them.

HEALTH PHYSICS

"Health Physics" was the wholesome and cheery name given to everything concerned with the protection of workers in the nuclear industry. Exposure to radiation can be *chronic* or *acute*, and the effects are different. It is first necessary to be able to estimate radiation "doses" numerically, and this is not an exact science, since there are so many variables. So far, we have had only a measure of the number of disintegrations per second, the curie, or 3.7×10^{10} dis/sec.

The rutherford, 10^6 dis/sec, is also used. But these are not really what we want. The biological effects of radiation depend on its *ionizing* power, since this represents molecular disruption, which is the root effect. Alpha and beta particles are charged, and when moving rapidly through matter, dislodge electrons. Alpha particles may knock other light nuclei into motion in collisions, and these knock-ons may also ionized. In air, every ion pair, electron plus ion, requires about 34 eV of energy. Alpha rays have ranges of a few centimeters in air, beta rays of a few metres. A sheet of paper stops both. Gamma rays may eject electrons from atoms, disappearing in the process, or may collide with electrons, knocking them on while being deflected (Compton effect), or if they are more energetic than 1.05 MeV, may produce an electron-positron pair. The fast electrons then cause ionization. The roentgen is the amount of × or gamma radiation that, in air, produces 1 esu of charge of either sign per cubic centimeter. (The electronic charge is 4.80×10^{-10} esu.) This is the basis of dose measurement.

For the effect of radiation in soft tissue, the *rad*, which is the amount of radiation depositing 100 erg per gram of tissue, was defined. It happens that this is just about equal to 1 roentgen. The relative biological efficiency, RBE, of a radiation is the ratio of the effect that a dose of X-rays of 200 kV energy would have divided by the dose of a particular form of radiation in rads. This expresses the fact that the same deposit of energy will do different amounts of damage in different tissues. Finally, the dose in rem (roentgen-equivalent-man) is equal to the dose in rads times the RBE. This procedure at least allows an estimate of the biological effects of various forms of radiation. Doses in rem are additive. Neutrons and gamma rays are the radiations of major concern in health physics. Fast neutrons have an RBE of 10, thermal neutrons 2.5, and gamma rays 1.0. Alpha particles, with an RBE of 10, would be very dangerous, but they are easily stopped, and it is almost impossible to get an alpha-ray dose unless the radioactive substance is inside the body.

An acute whole-body dose of 25 rem and under has no clinical effect. This once led to the belief that small doses were not injurious, but this is not true for chronic exposure. Greater doses may produce blood changes, but not until the dose exceeds 100 rem will there be a chance of vomiting and fatigue, with recovery in a few weeks. Over 300 rem the effects are serious, with hemorrhage, infection and hair loss, and recovery taking up to a year. Above 600 rem, the dose is usually fatal within a month or so, but a few may survive with serious effects.

Chronic low-level doses do not produce these dramatic responses, but may result in cancer, genetic damage, and similar effects. For people occupationally exposed to radiation, the chronic dose should not exceed 3 rem in 13 weeks, or 12 rem per year. The total dose should not exceed 5(N - 18) rem, where N is the age of the person in years. These were the levels accepted in the 1960's—they may well be lower now. Note how very much smaller they are than the acute doses with clinical effects. For people not occupationally exposed, the guideline is 0.5 rem/year, and no more than 1.5 rem up to age 30. These levels are comparable to the levels of natural radiation, from cosmic rays, uranium in igneous rocks, and other sources. The cosmic radiation dose is greatest at high altitudes, and decreases rapidly approaching the surface. At high altitudes, it seems to be less than 25 mr (approximately rem) per day, so a maximum yearly dose would be about 9 rem. At the surface, the dose would be much smaller, perhaps close to the recommended 0.5 rem/year.

Radiation can be detected in very small amounts. The dosimeter carried by workers in the field is a pencil-like capacitor that is charged to a high voltage. If you look in the end, an electroscope can be seen showing that the central wire is charged. A rather small amount of ionizing radiation passing through the device will produce enough charge to discharge the capacitor, and the amount of discharge can be seen by observing the calibrated electroscope. There are, of course, more sophisticated devices that are charged and measured by a machine. The Geiger-Móller tube responds to individual events, usually gamma rays, though beta rays can enter if a thin window is provided. It is good for low levels of radiation. Any dangerous level of radiation will completely block a G-M counter, so if you hear it ticking, you are quite safe.

Because radiation is silent, and can be detected in minuscule amounts, people who have trouble in estimating risks are excessively frightened by it, causing uranophobia and other mental illnesses.

It is extremely difficult to receive a dangerous acute dose as long as you stay away from unshielded high-level sources, such as are used for metallography or sterilization. The irradiation of food leaves no radiation in the food whatsoever, and makes fewer chemical changes than ordinary exposure to the environment (but often decreases palatability), while eliminating some actual, dangerous risks from pathogens. Exposure to dental X-rays, for example, carries little risk to the patient.

A lead apron is used to drape the patient, when it is actually the operator that is in most need of protection. This is almost completely for patient reassurance, not need. In the 1940's, fluoroscopes were used in shoe stores to show the bones of the foot in footwear.

To get a bright image, rather strong X-rays were necessary, although the part of the body exposed was rather insensitive to radiation, and stray X-rays were probably well blocked by shielding.

Of course, an end has been put to this entertainment, but one wonders if there was any real hazard. Whatever the answer, a salesman exposed many times daily was more at risk than any customer who visited once a year.

While on the subject of risk assessment, it might be good to mention the threat of a radioactive "dirty bomb" here. As a weapon, this is probably one of the worst ideas to appear in a long time.

It would be even less productive than chemical agents, which are more effective, but still prove to be very poor weapons. The radioactive material would have to be collected inside the device (a nuclear weapon makes its own just when needed), and to produce sufficient unpleasantness the amount would probably kill whoever transported the weapon to its place of use.

Proper shielding would be very heavy and cumbersome, rather difficult to conceal. Whatever is in the device would then be distributed by an explosion unevenly over a rather small area, which could be quickly evacuated before more than a small acute dose was received by anyone. A dose of 50 rem would require a very large amount of any kind of radioactive material that is available.

Anyone messing with fission fragment waste would become very dead very rapidly. The contamination could easily be detected, and warnings posted. Dirty bombs are of the same order of effectiveness as the Italian terrorists who thought they would poison the water supply with potassium ferrocyanide. It sounds like cyanide, but it is not poisonous, and wouldn't poison anyone. Give the emergency personnel Geiger counters, and forget about dirty bombs.

Chapter 5

Carbon

Carbon is the lightest member of the IVA family of the periodic table, atomic number 6, electronic configuration $1s^2 2s^2 2p^2$. Its first ionization potential is 11.26V. Its naturally-occurring stable isotopes are of mass 12 (98.89%) and 13 (1.11%). C^{14} is produced by cosmic rays in the atmosphere at a constant rate, and decays to stable N^{14} by β^- emission with a half life of about 5730 years. A living sample of wood, say, has the equilibrium amount of C^{14}. When it dies, the radiocarbon decays steadily, and the ratio of C^{14} to C^{12} declines to half in 5730 years, and so on. By measuring this ratio, the time when the wood was living can be estimated. This method of dating was worked out by Willard Libby in 1947, and is called "carbon dating."

The atomic weight of $C^{12} = 12.0000$ was established by the IUPAC in 1961 as the standard of atomic weights. Previously, the naturally occurring isotopic mix of oxygen was taken as 16.0000. This makes the atomic weight of natural carbon 12.010.

Carbon is the 19th most abundant element in the earth's crust, at 0.027%. It is found as carbon dioxide, carbonate rocks, and living matter. It occurs in elemental form in small amounts, as graphite and diamond. The properties of these forms of carbon, which are very important, are discussed in separate sections. The name is derived from the Latin *carbo, carbonis*, "charcoal."

Like silicon, carbon does not form ionic bonds. Unlike silicon, it forms covalent bonds with other carbon atoms, which can share two, four or six electrons to make single, double or triple bonds, and also single or multiple bonds with nitrogen, oxygen, sulphur and phosphorus. Carbon single bonds are usually sp^3 hybrid bonds pointing in tetrahedral directions, as in silicon. The angle between these bonds is 109° 28'. It satisfies its valences with hydrogen rather than with oxygen, as silicon does. This means that carbon can form an extremely wide range of compounds so that it has a chemistry of its own, *organic chemistry*. Only such a rich treasury of compounds can support the demands of life, so carbon compounds are the basis of all life. These compounds will not be discussed in this article.

We will discuss the elemental forms of carbon and their applications. Carbon is also at the centre of the fuels that support the present economy of the world, and fuels will receive close attention. All the curious and unusual uses of carbon that I can find will receive more attention than is their due, as well as the modern topics of buckyballs and nanotubes. Continued reflection turns up traces of carbon in the most varied applications, and these will be reported as long as space is available. We will see once again how important structure is to properties. Carbon is not black, or transparent, or soft, or hard, or a conductor, or a nonconductor—it all depends on structure.

DIAMOND

Our word *diamond* comes from the Latin *adamas, adamantis*, which is itself a Greek word, *adamas, adamantos*, meaning in these languages "hard steel," steel that had been heated red-hot and quenched to give a hard, brittle edge or surface. The diamond was a rare gem in the classical Mediterranean, coming from exotic India, also a source of steel and other curiosities. Pliny refers to the diamond as "adamas" as expressing its hardness, the property which gives it its industrial importance. The classical world treasured colored gems more than the clear diamond or quartz, but at the present day the sparkle and artificial costliness of the diamond put it in first place by far. An octahedral diamond crystal is shown in the figure.

Until 1955, all diamonds were natural, though attempts had been made to create them since the beginning of the century. In that year, the long search for artificial diamonds was finally successful, and small industrial-quality diamonds were made from graphite through high temperatures and pressures. In 1970, larger, gem-quality diamonds were artificially made, but they are more costly than natural ones. In the next decade, Russian scientists discovered how to deposit diamond from the gas phase to make hard coatings. Artificial diamonds now command a good fraction of the market for industrial diamonds.

The diamond crystal is made from carbon tetrahedra, just as crystalline silicon is made from silicon tetrahedra. Pauling gives a free-fusion stereogram of the structure.

The single bonds are 0.154 nm long. It is a face-centered cubic crystal with lattice constant 0.356 nm, giving a density of 3.51 g/cc, a rather high value for a nonmetallic substance. The pure crystal is transparent, with index of refraction 2.465 at 397 nm, 2.427 at 527 nm, 2.417 at 589 nm, 2.408 at 670 nm, and 2.402 at 763 nm. This difference of 0.063 between UV and IR makes diamond three times as dispersive as a typical glass.

Diamond is transparent to X-rays, unlike glass, and can be recognized by this property. Diamond is usually luminescent in UV light, due to impurities that are otherwise not apparent. The high band gap, 5.33 eV, means that diamond should be transparent in the UV to 233 nm. Its specific heat is low, 1.449 cal/g-K, as is its thermal expansion, 0.9×10^{-6} per °C. Its electrical conductivity is extremely low,

2×10^{-21} μS/cm, but its thermal conductivity is high, 5.54 cal/s-cm-K.

This high thermal conductivity gives diamond a cold feel, like quartz, but unlike glass. The only substance comparable to diamond in hardness is cubic boron nitride (discovered in 1957). In the Mohs scale, it is 10, but is much harder than the next competitor, corundum (Al_2O_3) at 9. This makes diamond the ultimate abrasive, able to grind and cut all others. The cubic crystal of diamond cleaves perfectly perpendicular to a 3-fold axis, or parallel to the faces of an octahedron. At a bright red heat, about 870°C, diamond burns to carbon dioxide. In the absence of oxygen, diamond sublimes around 3500°C. A perfect diamond is a single giant molecule, held together by covalent bonds.

The industrial uses of diamond depend only on its hardness; its gem uses depend on its unusually large index of refraction and dispersion. About 80% of diamonds go to industrial uses, 20% to gemstones. A gemstone, or *cuttable* must be transparent, free of bubbles, inclusions or visible flaws, light in colour, and of sufficient size. A *fine industrial* diamond may be dark brown, but otherwise of gem quality. These diamonds are used for wire-drawing dies of the best quality. *Bort* represents 99% of industrial diamond. It may be black, have flaws or inclusions, and small in size, but will still be an excellent abrasive. Bort is used for diamond drills, and for diamond wheels. A typical drill diamond may be 20 to the carat. About 1% is composed of microcrystalline aggregates that are very hard and do not cleave, called *carbonado*, or the spherical aggregates called *ballas*, most of which comes from Brazil. Its density is 3.15-3.29 g/cc. The three principal industrial uses of diamond are as a general abrasive, for making drills, and as wire-drawing dies. In 1957, the price of industrial diamond was about $4 a carat.

The weight of a diamond is given in metric carats of 0.200 g. The traditional carat varied from 0.197 to 0.207 g in different places. The international carat of 1871 was 0.205 g, and the metric carat was adopted in the U. S. in 1913. The largest diamond ever found, the Cullinan (1905), weighed a little over 3000 ct. If a diamond were a sphere of radius r, then $r = 2.39C^{1/3}$ mm, where C is the weight in carats. A 1000 ct diamond would be about 2" in diameter, a 100 ct 1", a 10 ct 1/2". You can use this to estimate the weights of any diamonds you may have around. The largest diamond found in the U.S., in Pike County, Arkansas (1906) was 40.22 ct, and the Punch Jones diamond found in Petersburg, W.Va. in 1943 was 34.46 ct. A 100 ct diamond can be considered exceptionally large and valuable, used only in royal jewels. A 10 ct diamond is large enough to be vulgarly displayed and generally admired.

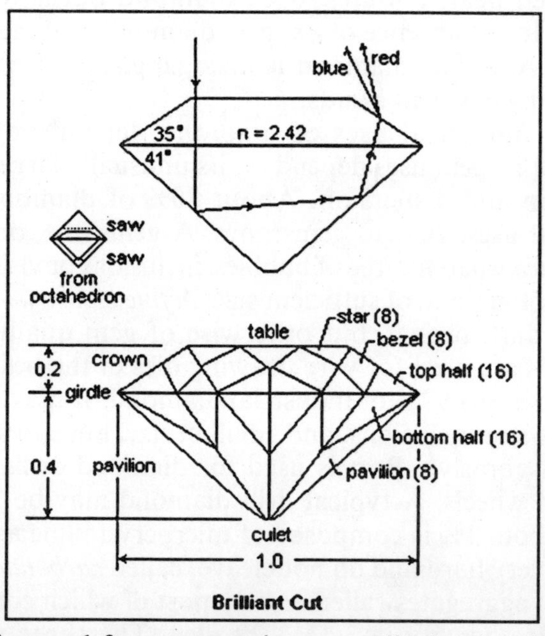

Brilliant Cut

A diamond for gem use is *cut* to display its clarity and brilliance. The most common cuts are derived from the *brilliant* cut, shown at the right, which came into use around 1700. It is based on an 8-fold symmetry, and the number of facets is a multiple of 8. The basic brilliant cut has $7 \times 8 = 56$ facets, plus the top face or *table* and small bottom face or *culet*, for 58 facets in all. Each facet has a name; typical names are shown. A major gem is cut from an octahedral crystal as shown, with its *girdle* corresponding to the widest part of the octahedron. About 50% of the crystal is used. The unused bits can be used to make smaller gems, called *melee*. None of the natural crystal is wasted. The height of the gem is about 0.6 of the diameter of the girdle.

The upper part, or *crown* is about half as high as the lower part, or *pavilion*. The purpose of the cut is to return light entering the crown approximately in the direction from which it came, without loss. About 17% of the incident light is reflected at the surface, much more than the usual 4% at a glass surface. White light is dispersed at the final refraction as the light exits the crown; the angular spread of the spectrum is about 2°. This causes colour at the boundaries of the reflected areas of light, which are broken up by the many facets. Colorless gem diamonds are preferred, but the rare blue diamonds are treasured. Most colored diamonds are yellow. Though this can be a beautiful colour, yellow diamonds are depreciated.

Since diamond is the hardest of all materials, it would seem impossible to cut and smooth the facets on a gem. There must be a trick, and this is it: the hardness of a diamond surface depends on direction. There are hard directions, and soft directions. Though there is not much difference, it is sufficient that the soft direction can be abraded by the hard direction. This must be taken into account even in sawing diamond by diamond, which is an arduous process. The facets of a gem are polished in their soft directions by diamond oriented in the hard direction. Cleaving is the only easy way to split a diamond, and here the slightest error can ruin the work. As might be evident, all this requires much skill and knowledge, so that diamond cutting is a special branch of gem work, distinguished from the less exalted skills of *gem cutters* and *lapidaries*, who have diamonds at their disposal and do not have to fret about hard and soft directions.

Natural diamonds were made 1-3 billion years ago in the earth's mantle, at a depth where the pressure was sufficient, say 55,000 atmospheres, and the temperature not too high, perhaps a little over 1000°C. These conditions obtain at a depth of about 150 km below continents, or 200 km below ocean crust. The rock there is peridotite, distinguished as Lherzolite or Harzburgite, composed of 60-90% olivine and 10-40% orthopyroxene, or else eclogite, 50% red garnet and 50% green clinopyroxene, with more iron and aluminium. A diamond begins as a tiny seed, perhaps some garnet, or iron sulphide or carbide. Carbon then grows around it in monomolecular layers, taking millions of years, which often results in well-formed euhedral crystals: octahedrons, cubes, cuboctahedrons, dodecahedrons with rhombohedral faces, or the twinned crystals called macles. The carbon in perioditic diamonds seems to be mantle carbon, while eclogitic diamonds contain carbon with an isotopic ratio typical of organic carbon. This carbon probably was pulled during plate subduction, since eclogitic diamonds were formed mainly in old subduction regions.

These diamonds have been carried to the surface in a few places by violent eruptions caused by the release of active fluids, probably mainly water, in the hot rocks of the upper mantle. The combination of these fluids and the perioditic and eclogitic rocks corrode their way toward the surface and regions of lower pressure, creating dikes and sills and eventually breaking out in hot, turbulent jets that reach the surface in carrot-shaped pipes called *diatremes*, carrying with them fragments of mantle rocks and rocks dislodged in their ascent, all in a hot, fluidized mass. Most of these eruptions have occurred in the last billion years, typically in the lower Palaeozoic (Devonian), but apparently none since the Cretaceous. The chaotic filling of a diatreme is the soft, easily eroded rock called *Kimberlite* or *Lamproite*, typically enriched in potassium, which is otherwise quite rare in such rocks. This is not the only way for diamonds to reach the surface, but it is the principal way, and all commercial diamonds come this way. These diatremes occur in stable craton regions of ancient rocks (because the diamonds were made beneath such regions when they were more active). They can be a half-mile in diameter, and are not rare.

Erosion of the kimberlite pipes frees the heavy diamonds from the matrix, and they join the bed load of streams that drain the area, winding up in alluvial deposits that may have been worked more than once. Diamonds, of course, are extremely durable. All ancient diamonds were found in alluvial deposits, as in India or Burma, and this remains a freqent means of discovering diamond deposits today. Few large diamonds occur in these deposits, but there is a large amount of medium and small crystals. The diamonds have an affinity for petroleum jelly, which easily entangles them. Alluvial diamonds were discovered in western Africa in the 19th century, notably in Congo, Ghana, Ivory Coast, Sierra Leone and other places around the Gulf of Guinea, and today support the economies and revolutions of these impecunious kleptocracies. All you need to mine alluvial diamonds is a shallow metal pan and some persistence.

Diamonds were discovered in southern Africa in 1867, when a 20 ct stone was picked up off the ground by a child. Soon after, the primary deposits of diamond were found in the vicinity of Kimberley, which gave its name to kimberlite. Although these deposits do not produce the majority of diamonds, they are impressive and famous for the scale and intensity of mining. Fresh kimberlite is also called blue earth. Weathered, it becomes yellow earth, from which the diamonds are easily separated. 1600 lb of blue earth may yield only 1/4 ct, 50 mg, of diamond. If they were not so valuable, they would be totally neglected. The Kimberley pipes are Cretaceous in age, among the youngest known.

The blue earth was mined originally in open pits, but then underground as the workings became deeper. At one time, the blue earth was allowed to weather so it was easier to crush and sort, but now the blue earth is processed at once, saving the cost of the inventory lying in the sun. The crushed blue earth is washed to separate the lighter gangue. The remaining 1% is then further enriched by *jigging*, as with gold, but the diamonds are not nearly as heavy as gold, so it is a more difficult process. Then the residue is *greased*, or allowed to move down a trough coated with petroleum jelly, to which the diamonds adhere much more strongly than the gangue does. This process was introduced in 1896. Then the greasy diamonds are *cleaned*, removing the petroleum jelly and leaving the product to be graded.

For many years, South Africa had a near-monopoly of diamonds, because of the scale of their operations. However, recent discoveries have pushed South Africa out of the limelight. Most of the industrial diamonds, perhaps 70%, used to come from the southern Congo. The Congo deposits are in part due to a Triassic pipe at Bakwanga, whose erosion left diamonds in the Karroo gravels. Botswana is now a larger diamond producer than South Africa. In 1960, Devonian kimberlite pipes were found at Zarnitsa, in the Yakut A.S.S.R., now the Republic of Sakha, in deepest Siberia. These deposits, which include gem-grade diamonds, have made Russia one of the largest diamond producers. Probably the most significant new discovery came in 1981 in far northern Western Australia, at the Argyle mines in the far northeastern corner near Wyndham, and at Ellendale near Fitzroy Crossing. Exploration began in 1969 after 9 diamonds were found in the Leonard River. Full scale exploitation began in December 1985. The mainly eclogitic diamonds are in lamproite, and include rare and valuable pink diamonds. Of the Argyle diamonds, 55% are industrial, 45% near-gem, and 5% gem quality, of which 95% are brown, 4% colorless or grey, and only 1% the valuable pink or red. There has been some effort to sell the brown diamonds by giving them attractive names.

Near Murfreesboro, Arkansas, J. W. Huddleston picked up some diamonds in Prairie Creek in 1896. One was cut to a 3.0 ct white diamond, and the other to a 1.5 ct yellow, in New York, which made the site, Crater of Diamonds, where the diamonds were found in a lamproite pipe. The largest found, in 1906, was the Uncle Sam, 40 ct, referenced above. This was never a commercial mine, and is now a state park, where you can pan for diamonds for a small fee. The source of the Punch Jones diamond found in West Virginia seems to be unknown, and no more have turned up.

Kimberlite or lamproite pipes from late Proterozoic to early Devonian have been found in metamorphosed sediments and volcanic rocks of early Proterozoic age in the State Line field between Laramie, Wyoming and Ft. Collins, Colorado, and in the Iron Mountain field northeast of Laramie. At least 40 diatremes have been located. The search for these began in 1975, and a 28.3 ct gem diamond has already been found. Serious mining was supposed to have begun in 1996.

Some time ago, a diamond was found in glacial till in Wisconsin, that had been carried south by Pleistocene continental glaciers from somewhere in Canada. In 1981 some diamonds, or diamond indicators, had been found in the MacKenzie River, so there was some prospecting by large diamond-mining concerns, but nothing was found, and they sold out to one of their contractors, Chuck Fipke. Fipke finally found what he knew was there in 1985, under Lac de Gras, 300 miles northeast of Yellowknife. This finding was made with the scientific assistance of Professor John Gurney of Cape Town University in South Africa. 120 pipes were found, 5 minable, in the Ekati area. Full operation began in October 1998, and Canada became a major player in the diamond game. At Diavik, 41 kimberlite pipes were found, 13 of which contained diamonds, and 4 were rich.

The indicators of diamond are minerals that are normally quite rare at the suface, including garnet, chromite, ilmenite, clinopyroxene, olivine and zircon. The last is the most persistent in sands, and is an excellent indicator that there are kimberlite pipes upstream. Most of these minerals are deeply colored, like red garnet, black chromite, and green clinopyroxene and olivine, and so are easily picked out. They contrast greatly with the usual silica, mica and feldspar. Before the recent entry of Canada, 92% of all diamonds came from Australia, Russia, Botswana, Congo and South Africa.

Brazil also has several diamond regions, but has not made as great an impact on diamond production. Curiously, diamonds have recently (1999) been found in French Guiana, not in kimberlite pipes, but in a greenstone schist, in a mineral called komatiite, at longitude 53°W, latitude 4°N. The Dachine greenstone belt is the remains of the roots of a Proterozoic island arc, where there was once subduction, and the schists were made by the metamorphosis of volcanic material in the

Amazonian orogeny (perhaps when South America collided with North America in the Ordovician). This rock, now exposed by erosion, contains only small diamonds. It is interesting that at the time, Africa was in close contact, and this area was opposite the areas of C&ocaret;te d'Ivoire and neighboring countries where alluvial diamonds are now found.

GRAPHITE

The name *graphite* comes from the Greek *grapho*, "I write" by allusion to its use in pencils. Graphite indeed leaves a black mark, but it is too soft unless mixed with clay, pressed into thin rods and consolidated by heat. These are called "leads" because soft lead leaves a similar (but not so good) dark mark. The word *plumbago* for graphite alludes specifically to lead. In ancient times it was probably regarded as yet another form of lead, a more concentrated *plumbum nigrum*. Graphite pencils do not seem to have been used at all commonly in the ancient world. A stylus on hard wax, or ink on papyrus, were sufficient for most purposes. A specimen of graphite is shown in the figure.

Fig: Graphite

Graphite is an allotropic form of carbon, as different from diamond as could be imagined, except for one thing. It, too, is formed by covalent bonding of carbon atoms, and is just as resistant to chemical attack as diamond, and laughs equally well at high temperatures. Graphite does not melt, but sublimes at about 3500°C. Crystalline graphite comes in two forms, α-graphite, which is hexagonal in symmetry, and β-graphite, which is rhombohedral. Most of the time when carbon is reduced to an element, it associates as some form of graphite, though usually not as obviously crystalline graphite. These other forms are often called *amorphous graphite*, which is good enough for engineers, but all such material is crystalline on a small scale. Graphite is the reason it is so hard to make diamond. Carbon would much rather be graphite under normal conditions. It is more stable than diamond, but diamond never alters to graphite.

Structure of Graphite

The structure of &alpha-graphite is shown at the right. Pauling gives a free-fusion stereogram of the structure. A carbon atom is located at each vertex of a hexagon, and each line joining two vertices represents a single bond with two electrons. These bonds are 0.142 nm long, intermediate between the 0.154 nm of a single bond and 0.133 nm of a double bond. Each circle represents two delocalized electrons that distribute themselves evenly around the hexagon. To each hexagon corresponds two (6/3) carbon atoms, 6 (12/2) electrons in localized bonds, and 2 delocalized electrons, which accounts for the four valence electrons of each carbon. The delocalized electrons are responsible for the great stability of graphite. Each sheet of hexagons is a separate molecule, a natural colloid that is all surface. The sheets are held together by van der Waals forces that are much weaker than the covalent bonds. The fluctuating charges in one sheet induce opposite charges in the other sheet, so the sheets attract. In the α-graphite, they are 0.340 nm apart. Heat and electricity flow easily in a sheet, with difficulty normal to the sheets. The resistivity normal to the sheets is 10,000 times the resistivity in a sheet. A sheet can stick to another plane surface, as well as adsorb molecules and exhibit all the usual surface phenomena. The uses of graphite as a lubricant and as an adsorbent depend on its surface.

The single bonds in graphite are formed by $sp^{<2}$ hybrid orbitals that lie in a plane at angles of 120°. The remaining electrons are in a p_z orbital, half above the plane, half below. These orbitals overlap to form the circular orbitals above and below the plane that accommodate the delocalized electrons to give graphite considerable “resonance” energy.

Although diamond is a semiconductor, graphite is classed as a *semimetal.* Its valence band is filled, and its conduction band is empty, at absolute zero, but the energy gap $E_g = 0$. Therefore, there are always mobile electrons in its conduction band, and mobile holes in its valence band, at nonzero temperature. These charge carriers are not responsible for its bonding, as in a metal. They are, however, responsible for the semimetallic lustre of graphite. The black of graphite is due to its fine division, usually colloidal, which absorbs light effectively.

One way to make a colloidal graphite is to burn things that make a smoky flame, like resins, and to catch the soot on a cool surface. The soot is called *lampblack*. In Latin, this material was called *fuligo, fuliginis*, and was used to make black paints and the important substance *atramentum librarium*, or ink. To make ink, you ground fuliginem in a mortar with some gum, *gummi*, perhaps from the acacia, and a little water. The colloidal graphite was then peptized by the gum, forming a stable sol that was the ink. This recipe was discovered in Egypt when writing on paper was invented, and is now known as India ink. The formula is given in Vitruvius. Ink, paper and writing all came to the west from Egypt. Ink was also discovered in China, where gelatin was used instead of gum. Writing in a different fashion also came from China, with paper made from textile fibres. This excellent ink is still used today in drafting, a survival of a very ancient technique.

Graphite is found in nature in three forms, known in the trade as *crystalline flake*, *crystalline lump*, and *amorphous lump*. Crystalline flake graphite occurs as flakes in the metamorphic rock mica schist. The world's best deposit is in Madagascar, though it is widespread. In Madagascar, 13 tons of schist yield 1 ton of 90% graphite flake. This is some of the most desirable graphite in trade, called "crucible grade" and also used in lubricants, where all abrasive impurities must be absent. A primary use of graphite is in making refractory crucibles and other objects that must stand extreme heat. Graphite vanes were used to steer the V-2 rockets in world war II.

Crystalline lump occurs in veins associated with intrusive pegmatites in Ceylon. The graphite veins radiate from the pegmatite inclusion. The source of the graphite is probably the reduction of the common volcanic gas carbon dioxide. Amorphous lump graphite is produced by the alteration of coal by nearby granite intrusions, which cooks out the volatile ingredients in a kind of natural coke oven. In fact, there is a continuous progression through anthracite coal to graphite. The structure of coal is different from that of graphite, however. These graphites can be used to add carbon to steel, and other uses that do not require high purity. The largest veins of amorphous graphite are in Sonora, Mexico, bedded in a triassic sandstone. These were indubitably coal deposits cooked by nearby intrusions. Similar deposits occur in Korea, which supplies most of the world's lump graphite. New Mexico and Rhode Island also have vein graphite deposits.

Dr Acheson found how to make good graphite in an electric furnace. It seems to be competitive with natural graphite, and is generally purer. Graphite can be peptized by tannin to make an aqueous sol called *aquadag*, and a dispersion in oil called *oildag*. The "dag" syllable seems to be a term for *deflocculated graphite*, a product of the same company. Aquadag can be used to deposit a conducting graphite layer on the interior of a cathode-ray tube to complete the electron circuit. Graphite is used for the cathode in ammonium-zinc and alkaline electrical cells because of its inertness and conductivity, and as carbon arc electrodes because of its high sublimation temperature.

Graphite is a good lubricant when in fine flakes, even when dry. This is not because the graphite layers slide easily on each other, because they do not, but stick persistently. When a layer has adsorbed neutral molecules on both sides, however, it does become slippery. It tends to plate metal surfaces, so it does not rub off.

Carbon black is the colloidal graphite used in rubber products, and is identical to lampblack. It is usually made by burning methane in limited air. Rubber is very susceptible to abrasive wear, and tires would last a very short time unless the rubber were mixed with large quantities of carbon black, which gives it wear resistance. A chemical reaction is probably involved, part of the vulcanization process. The general black colour of rubber products is evidence of the use of carbon black. Pure rubber is a pasty whitish colour.

Thomas Edison found that the resistance of carbon granules depended on the pressure to which they were subjected. By connecting an acoustic diaphragm mechanically to a button of carbon granules, the resistance could be made to vary in accordance to the acoustic pressure, so that when a current was passed through the button an amplified replica of the acoustic signal was produced. This gave a much stronger signal than the Bell type of metallic diaphragm, permanent-magnet microphone, and made long-distance telephony possible. Edison, as is well-known, also used carbon filaments in the first incandescent lamps. The search for the best material to carbonize is a fascinating story. Bamboo fibres turned out to be the best.

Carbon paper is another useful invention largely ignored today (though carbon paper can still be purchased). It uses a film of carbon black on thin paper for the purpose of making multiple copies simultaneously by being struck with a type head of a typewriter, or embossed with a stylus held in the hand. An early use was making multiple copies all at one time of telegraphic railway train orders, messages that had to be precisely correct, so they were guaranteed to be identical. In the business world, a typist could make a file copy at the same time as the original letter, or make a copy for another recipient. There had been ways of making multiple copies before carbon paper, but carbon paper was far superior to them.

Black powder is black because it contains graphite from charcoal. Gunpowder is made from potassium nitrate, charcoal and sulphur very intimately mixed by grinding when wet. This is a large subject that cannot be adequately treated here, except to say that making gunpowder is somewhat more difficult than would be guessed, and there are many tricks involved. Gunpowder actually burns very rapidly rather than exploding, and this is what is desired in a firearm. Gunpowder Creek in Wyoming was so-named by the Cheyennes because of the coal that crops out on its banks. Gunpowder was the only black powder that they knew.

OXIDES

Carbon forms three oxides: carbon monoxide, CO; the dioxide, CO_2; and the unimportant but curious carbon suboxide, C_3O_2. The suboxide is a colorless gas condensing to a liquid at 7°C. In water, it hydrolyzes to malonic acid, HOOC-CH-COOH, of which it is the anhydride. Its structure is linear, O=C=C=C=O.

+1 -1	0 0	-1 +1	
:C–O:	:C=O:	:C≡O:	0.113 nm
+1 -1	0 0	-1 +1	
:O≡C–O:	:O=C=O:	:O–C≡O:	0.116 nm

Resonance Structures with Format Charges

The Lewis structures of CO and CO_2 are shown at the

right. Each line represents two electrons in a covalent bond; assign one electron to each atom joined by the bond. The dots represent electrons belonging to a particular atom, here the oxygens. Let the oxygen have a formal charge of +6, the carbon +4, each electron -1. Then the formal charges are as shown. The centre structure is nonpolar, while the outside structures are polar in opposite directions. The state for the molecule as a whole is a linear combination of the three structures. The ability to choose the relative amounts of the structures allows the formation of a state of lowest energy. The is rather inappropriately called "resonance" in chemistry. There is no resonance involved, just the freedom to adjust the relative amounts. The amounts of the two polar structures are equal, so the molecule is nonpolar overall. The resultant lengths of the bonds are shown at the right. The possibility of "resonance" makes these molecules especially stable.

Carbon monoxide is a very stable diatomic molecule. It is a colorless, odorless gas at STP. Its molecular weight of 28 makes it just a tad lighter than air. It burns to carbon dioxide in the presence of oxygen, releasing a good deal of heat (67,700 cal/mole), and reduces iron oxide to the metal, becoming carbon dioxide. Banefully, it combines tightly with hemoglobin in the blood, taking it out of action and suffocating from within. For this reason, it is a deadly poison in air that is breathed. One part in 100 is rapidly and irreversibly fatal. One part in 500 brings unconsciousness in 30 minutes, death in three hours. Otherwise, it is just fine. Carbon monoxide is created when carbon is burned in a deficiency of oxygen, as in charcoal fires, internal combustion engines, poorly drafted and vented heaters and furnaces, fuel gases, and in methane explosions in coal mines. The symptoms of inhalation of small quantities are headache and nausea. It is treated by inhalation of oxygen containing 5% carbon dioxide. Carbon monoxide can be detected by the blackening of paper soaked in palladous chloride solution, or in smaller quantities by electronic sensors.

CO is produced endogeneously in the body in the breakdown of the heme molecule when an erythrocyte's 120-day life is over. If it were not for a molecular adaptation in myoglobin and hemoglobin, the proteins that carry oxygen in the muscles and blood, this would make enough CO to block the globins fatally. As it is, about 1% of the globin is blocked by CO in normal circumstances. A free heme is 25,000 times as sensitive to CO as it is to O_2, but the globins only 200 times more sensitive. Life is complicated.

Carbon monoxide can be removed from internal combustion engine exhaust by adjustment of the engine, but this increases the NO_x emissions, which are even more poisonous. No hay lonche gratuito. Catalytic converters mainly catch unburned hydrocarbon fragments, and don't do much with CO and NO_x.

Carbon monoxide can be catalytically hydrogenated to make methanol (wood alcohol), which is a violent poison when taken internally, but nice enough otherwise, and a good solvent. With chlorine, it forms carbonyl chloride, $COCl_2$, also known as *phosgene*, a rather unsuccessful war gas, but a useful chemical intermediary in making dyes. The patented compound *Hopcalite* (Johns HOPkins and U. of CALifornia) oxidizes CO to CO_2 in gas masks. It is a mixture of manganese dioxide with copper and cobalt oxides, and must not be exposed to the slightest amount of moisture.

In coal mining, CO is called *black damp*, and occurred after there had been an underground fire or explosion. Canaries were used to detect it, since the birds were more sensitive than the miners to its effects. The safety lamps of Davy and Stephenson were used to detect the presence of methane, or *fire damp*, (by an enhanced flame) without setting it off. CO also had an effect on the flame. CO_2 was called *choke damp*, because it suffocated. The word "damp" here is cognate to German "dampf" and simply means a gas.

Carbon monoxide forms complexes called *carbonyls* with metals, like nickel tetracarbonyl, $Ni(CO)_4$. The carbons are tetrahedrally bound to the nickel, with the oxygens on the outside. There is "resonance" between the uncharged structure and structures with various charges on the carbon and oxygen that bind the molecule. Without "resonance" there would be no binding at all. Nickel carbonyl is a volatile liquid melting at -25°C and boiling at 43°C. It is formed by passing CO over warm Ni powder, then it is condensed and distilled for purity. The pure carbonyl is then pyrolytically decomposed at 150°C to deposit pure nickel. Carbonyls are usually inflammable clear liquids easily decomposed by heat.

Carbon dioxide is a very stable linear triatomic molecule, generally represented by O=C=O. Its molecular weight of 44 shows that it is heavier than air. It is colorless, odorless and nonpoisonous. It does stimulate breathing, however. The critical temperature is 31.1°C, so it is easily liquefied by pressure alone, and is supplied in steel tanks. The liquid freezes under 5.1 atm pressure at -56.6°C to a white solid, *dry ice*. This is the *triple point* at which gaseous, liquid and solid can coexist.

At 760 mm CO_2 pressure, dry ice sublimates at -78.5°C, and can freeze ice cream harder than a rock. At lower pressures, the sublimation temperature is even lower. It provides pure cold, without messy liquids. It is a normal minor constituent of the atmosphere, about 350 ppm by volume. Nevertheless, this small amount supplies all the food in the world for all living things, and is part of a complex natural cycle. Carbon dioxide, incidentally, dissolves in rubber and diffuses through rubber faster than helium or hydrogen does.

Carbon dioxide is quite soluble in water. At 1 atm, 1 litre of water dissoves 1.713 litres of dioxide, making a pleasingly acidic, effervescent solution. The equilibrium concentration in the solution is proportional to the carbon dioxide pressure. Soda water is charged with carbon dioxide at 4 to 5 atm pressure (50-65 psi), which slowly evaporates if the solution is open to the air, more rapidly if the solution is agitated. Supersaturation is easily observed when an opened can of soda is poured into a glass, and gas is then copiously evolved. This supersaturation is very desirable in a beverage. All rainfall is slightly acidic, even when absolutely pristine. Water in equilibrium with the usual atmospheric content of CO_2 has pH 5.7. The carbonate ion is formed by the action of the water: $CO_2 + H_2O$ "! H_2CO_3 "! $H^+ + HCO_3^-$ "! $2H^+ + CO_3^{--}$. This series of equilibria show that the solution contains a number of molecular constituents. The ionization of water means that the hydroxyl ion concentration is very low.

The carbonate ion has a threefold axis of symmetry, and is very stable, forming a wide variety of important salts. H_2CO_3 is called *carbonic acid.* Its true dissociation constant is 4.4×10^{-4}, so it is a fairly strong acid. However, the concentration of H_2CO_2 is so low that the actual H^+ concentration is also small. The apparent dissociation constant $[H^+][HCO_3^-]/[CO_2]$ is only 4.2×10^{-7}, so it is effectively a weak acid. Although real enough in solution, it cannot be isolated out of water as a crystalline substance, and always exists in equilibrium with dissolved CO_2. Carbonic acid seems to speed alcohol from the stomach to the blood stream, an interesting effect of carbonated drinks. The dissociation constant of bicarbonate to carbonate is 4.8×10^{-11}, and of water 1×10^{-14}. At a carbon dioxide pressure of 2.9 atm, the pH of the water solution is 3.4; at higher pressures it does not go much below 3.3.

Calcium carbonate is a very important carbonate, found in nature in limestones. Its rhombohedral crystals are really just carbonate ions, all pointing in the same direction, made electrically neutral by Ca^{++} ions at strategic locations. Clear crystals are called *calcite.* There is also an orthorhombic variety, called *aragonite.* It is quite insoluble, but calcium bicarbonate, $Ca(HCO_3)_2$, is not. When water that has absorbed some carbon dioxide trickles across it, the bicarbonate ions pick up some calcium from the carbonate, and wash it away in solution. This is an important weathering process that explains the weathering of limestones in a moist region along cracks and passages, producing great and small caves, underground streams, and all the wonder of Karst topography. (Karst is a limestone area in the Balkans that is an example of this.) This is part of a subcycle of the great carbon cycle in which animals in the sea make shells of calcium carbonate, which becomes limestone, which is raised above sea level, which is rained upon, and which returns to the sea as bicarbonate. Water containing calcium bicarbonate is the classic "temporary" hard water, since boiling deposits the carbonate as scale. The calcium forms the familiar insoluble scums with soap, so "softening" water means removing Ca^{++} ions, and Mg, Fe and Mn ions as well. "Permanent" hard water contains anions such as SO_4^{--} that will not precipitate the cations. In this case, the addition of carbonate is necessary, perhaps as washing soda, Na_2CO_3.

Carbon dioxide in water also weathers silicates to calcium bicarbonate and silicic acid, both of which are soluble. Carbon dioxide is released when silica and calcium carbonate react in boiling water to make calcium silicate and carbon dioxide. This is the origin of the "geyserite" produced around hot springs and geysers, and is the reverse of the first reaction. The weathering of feldspars to kaolin is one way carbon dioxide is removed from the atmosphere. Another is the fixing of carbon dioxide by organisms to calcium carbonate. Photosynthesis also removes carbon dioxide in large amounts, while certain bacteria use carbon dioxide in chemosynthesis. Most photosynthesis occurs in the surface layers of the ocean, it should be remembered, not by the more visible green land plants. The ocean contains 20 to 30 times as much carbon dioxide as the atmosphere, and is dominant in the carbon dioxide cycle. All this is very difficult to estimate quantitatively. It is no wonder that computer programs may seem to give observed variations; such programs are constrained to do so.

Carbon dioxide can be made by "burning" limestone. With heat, limestone decomposes according to $CaCO_3 \rightarrow CaO + CO_2$. This is a classic way of making the dioxide. CO_2 is a by-product of lime burning, cement making and brewing. Lime, CaO, is "slaked" in water to form the hydroxide: $CaO + H_2O \rightarrow Ca(OH)_2$. Mixed with sand and water, the hydroxide forms *mortar*. The hydroxide first crystallizes rapidly, and the morter sets. Then it gradually reacts with the carbon dioxide in the air to return to the carbonate, and hardens even more. Carbon dioxide is also made in fermentation, when enzymes catalyze the reaction turning sugar into ethyl alcohol and carbon dioxide. This is the major source of industrial CO_2. Yeast contains the enzyme zymase to accomplish this creditable task. Unlike methyl alcohol, ethyl alcohol is a good drink that is the source of much government revenue and a boon to the human race. Beer is an exceptionally healthy beverage that we should consume more of.

The carbon dioxide industry has been well developed. It supplies liquid carbon dioxide in strong steel cylinders, and solid carbon dioxide in compressed bricks. The liquefaction of gases was introduced by Faraday, while Thilorier discovered solid CO_2 in 1835. Solid CO_2 can be made by direct free expansion of liquid CO_2 to form CO_2 snow, which is then compressed in bricks. If no other source of the gas is available, it can be made by burning coke in air. The flue gas is scrubbed over limestone chips to remove SO_2 as calcium bisulfite, then the CO_2 is absorbed by an alkaline solution (Na_2CO_3, for example, called "lye") in a coke-filled tower and released in purified form by boiling the lye. The gas is then dried and compressed, and condensed to a liquid, eliminating the nitrogen. Some sources of gas, such as fermentation, may introduce unwanted odors and impurities. Cylinders holding 50 or 20 lbs of the liquid, or smaller cylinders for certain applications, are filled to about 60%. If they are overfilled, expansion of the liquid can produce dangerous stresses.

Liquid carbon dioxide has had many uses. As a compact source of power, carbon dioxide can be used in place of compressed air. Before electricity was widely available, it was used to operate some railway signals in isolated locations. Similarly, it was used to operate bell buoys for navigation. It was used in coal mining as an explosive ("Cardox"), by heating it in a closed cylinder until a shearing disc was ruptured and the high pressure (10,000 - 30,000 psi) released fractured the coal. It was used as a safe refrigerant until displaced by Freons. Now that Freons are out of favour, it may again be used for this purpose. The greatest use of liquid carbon dioxide is in the carbonation of beverages, and in raising beer. It not only can provide the power to lift beer from the cellar to the tap, but preserves the carbonation in doing so. Dry ice is used where other methods of refrigeration would be heavy or inconvenient, particularly in the distribution of ice cream. It provides an inert atmosphere for processes at moderate temperatures (at high temperatures, it dissociates to $CO + O_2$, so cannot be used for welding).

Carbon dioxide is found naturally as a major consitituent of volcanic gases, and in caves, mines and wells. At Price, UT, a well produced 98% CO_2 at 760 psi. Some oil wells in western Colorado produced an emulsion of CO_2 and crude oil, and were called "ice cream wells." It occurs in many naturally carbonated spring waters, as at Saratoga Springs, NY (in equilibrium with 21 psi CO_2), as well as at famous locations such as Seltzer, Spa, Vichy and Apollinaris. The CO_2 in these waters helps them to dissolve minerals, hence they are "mineral" waters. At the Grotta del Cane near Naples, the gas is 70% CO_2, 24% N_2 and 6% O_2, and is definitely fatal.

The Stygian Caves in Yellowstone, and the Lachersee in Germany, also are associated with carbon dioxide. Fatal properties of these gases are associated more with the lack of oxygen than the presence of carbon dioxide. Compositions of 88% N_2, 10.6% CO_2 and 1.4% O_2 (in a manhole) and 48.5% N_2, 38% CO_2 and 13.5% O_2 (in a silo filled with green corn) have proved fatal. On the other hand, a mixture of 5% CO_2 and 95% O_2, called "Carbogen," is used to stimulate breathing in CO poisoning and in termination of anaesthetics. The greatest stimulation of breathing occurs at about 7%, but concentrations over 25% or 30% tend to terminate breathing. Carbon dioxide plays a very important role in regulating breathing in normal concentrations. A human exhales about 900g of CO_2 per day. Since there are about 6×10^9 humans at present, this is not a negligible contribution to greenhouse gases.

Sodium carbonate, Na_2CO_3, and sodium bicarbonate, $NaHCO_3$, form clear hydrated crystals, but are more commonly known as white powders. They are the household alkalis, as vinegar is the household acid. Sodium carbonate decahydrate is called *washing soda*, since it is used to precipitate calcium carbonate to soften water for laundry, and sodium bicarbonate is called *baking soda*, since it is a source of CO_2 to leaven baked goods. When dissolved in water,

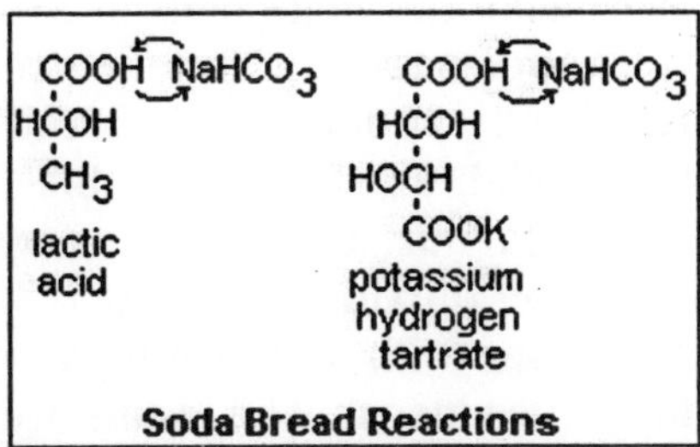

Soda Bread Reactions

The carbonate ion ties up all the H^+ ion in scarcely-ionized molecules, leaving the OH^- to rule. Sodium carbonate produces a strongly alkaline solution that irritates the skin and is very unpleasant to drink, but bicarbonate is an excellent remedy for excess stomach acid. A half-teaspoon in 5 ounces of water will cure indigestion at once with a satisfying belch, at little cost. This is the reason more expensive medications are touted, and bicarbonate ignored.

Sodium carbonate and bicarbonate are found naturally as the mineral *trona*, but much is produced artificially, since natural deposits are uncommon and often impure. It is found in evaporites in arid regions, and in fossil evaporites underground. The LeBlanc and Solvay processes have been used to manufacture the carbonate, both of which begin with salt, and the carbonate supplied by limestone. Pure sodium carbonate is a prime ingredient of soda glass. It is also used in making paper. Boiled with slaked lime, and then with fats, it makes soap. Soap dissociates in water to a molecule that has a polar end, and a nonpolar fatty end. It *peptizes* oil to make *micelles* with the polar ends on the surface that disperse in water and can be rinsed away.

Baked goods are foams of CO_2 in a gel of glutin, making grains easily digestible and pleasant. The CO_2 can be provided by yeast, which not only ferments the sugar, but makes its own sugar from the starch in the flour. It is usual to add some sugar to make the job easier. The alcohol evaporates in the heat, incidentally. The gas can also be provided by sodium bicarbonate, and then we have *soda* bread in place of *yeast* bread. The American "biscuit" and the Scottish "scone" are soda bread. American "corn bread" is soda bread carrying corn meal, which lacks the ingredient necessary to make a foam, but provides sweetness. Without the strong wheat flour, it would be just ghastly "pone." Soda bread can be just as delicious as yeast bread, and is less trouble. The only thing missing is the yeasty flavour. Two reactions are generally used to produce the carbon dioxide. The first is with lactic acid, which is present in sour milk, buttermilk, and milk "soured" on the spot with a little vinegar. The results are sodium lactate and gas. Another, using dry ingredients that activate only when moistened, is with potassium hydrogen tartrate. The dry mixture is called *baking powder*. The hydrogen is exchanged for the sodium, and we have sodium potassium tartrate, or Rochelle salt, and gas. The gas results from the dissociation of the carbonic acid, as usual. The reactions are illustrated with structural formulas in the diagram. In each case, a hydrogen and a sodium change places. These are very familiar and useful reactions that every cook should understand.

The old red and brass soda-acid fire extinguishers that used to hang in school corridors had a tank full of sodium bicarbonate solution, and a bottle of sulphuric acid with a loose stopper suspended at the top. When taken down and inverted, copious carbon dioxide was evolved, which pressurized the closed top (which was now the bottom) and squirted water out of the nozzle to be directed at the fire. The CO_2 helped to blanket the fire as well, but inefficiently. These were cheap and effective, but being cheap, were eliminated in favour of more advanced and expensive pressurized extinguishers that leak propellant and are duds when an emergency finally arises. Schools don't have too many oil fires, but when one does happen, a water extinguisher does not work. Foam extinguishers are much better, especially when chemically propelled in the old fashion and not by pressurized nitrogen.

As carbon dioxide is produced by fermentation, it is also produced in the formation of coal and petroleum from organic matter, probably with the aid of bacteria. This carbon dioxide usually escapes, but is occasionally found where it has been trapped. In Wyoming, Colorado and New Mexico, there are many carbon dioxide wells that produce the gas from deep porous rocks. This is probably gas that has never been in the atmosphere. It is used for pressurizing natural gas reservoirs, and in secondary recovery of petroleum, where it also has a solvent effect.

Carbon dioxide is called a "greenhouse gas" because it absorbs and emits infrared radiation. This slows the radiation of heat by the atmosphere into space in the same way that the glass windows of a greenhouse do. The visible light with its energy can get in, but the infrared radiated by the things it heats inside cannot get out as easily through the glass, so the greenhouse stays warm in the winter without the expense of fuel. The more carbon dioxide, the larger the effect. Carbon dioxide is not the only greenhouse gas, and is actually a minor one.

Water is the most important greenhouse gas. Methane and nitrous oxide are more effective than carbon dioxide, but present in smaller amounts. If it were not for the greenhouse effect, the surface of the earth would be as cold and inhospitable as the surface of Mars (almost). Stick your hand out the window the next time you are flying at 35,000 feet to see what the temperature is like out there. It is cold enough to freeze the spherical objects off a copper-alloy primate. Long-range climatology is a delicate balance that we do not know enough about to say anything for certain. There must either be global warming or global cooling (the only choices) and we seem to be stuck with the warming. It has been much warmer, and much cooler, in the past, mainly warmer. Causes we do not understand. What we *can* do is agitate for more research support.

FOSSIL FUELS

What are called *fossil fuels* are naturally-occurring substances composed mainly of carbon and hydrogen. On combining with the oxygen of the air, they yield water, carbon dioxide and, most importantly, large amounts of heat. They may be solids, liquids or gases. All are found in the upper crust of the earth, but only in sedimentary rocks that have not been subjected to great heat or pressure. All are the products of former life, produced by photosynthesis from water and atmospheric carbon dioxide, with the input of radiant solar energy. Their supply is strictly limited, and compared with the current demand, small. This applies with special emphasis to the fossil fuels available at low cost.

Fossil fuels appear in three major forms. Carbonaceous deposits are mainly carbon, with very little hydrogen and other elements. These are represented by *coal.* Bituminous deposits include hydrogen as well as carbon, and are represented by natural gas (gaseous), crude oil (liquid) and bitumen, asphalt or tar (solid). They are called, collectively, *petroleum,* though this term is popularly applied specifically to crude oil. The third form is pyrobituminous deposits, which must be distilled before they can be used. This form includes oil shales, and special varieties of coal. All fossil fuels are the remains of living matter. Coals arise from forest plants the size of trees, whose structures are clearly recognized in the coal. No coal is earlier than Carboniferous, since there were no appropriate plants before then. The other fossil fuels are the remains of microscopic organisms, chiefly diatoms and similar sea plants, which have existed from the Ordovician to the present in very large numbers. This is shown by the existence of optically active substances in them, and the presence of porphyrins. All fossil fuels have been produced with the aid of anaerobic bacteria that extract oxygen and evolve methane. There is no possiblity of an inorganic origin of fossil fuels.

Most terrestrial carbon is in the form of carbonates in sedimentary rocks, or as dissolved carbon dioxide in the seas, about 4.6×10^{15} tons. Some 7.1×10^{12} tons is in the form of coal and similar substances, while 1.1×10^{12} tons is in soil and living matter. The amount of carbon in petroleum is negligible. Of the oxygen that is not in water, about 8.0×10^{15} tons is in sedimentary rocks, 1.2×10^{15} tons is in the atmosphere, but only 1.2×10^{13} tons is dissolved in the sea. Note that carbon dioxide is very soluble in water, while oxygen is not. These figures are from Holmes (1945). Originally, the carbon dioxide was all in the atmosphere, while the oxygen was all in water. Life processes, mainy photosynthesis, have been responsible for the current state of affairs.

Plants on land use sunlight to take the hydrogens from water to make hydrocarbons from atmospheric carbon dioxide, releasing the oxygen as a waste product. This is done by the active membranes in *chloroplasts*, which use *chlorophyll* to absorb light at long and short wavelengths to gather the necessary energy. Plants still respire as all modern life does, by oxidizing hydrocarbons in *mitochondria* which take in oxygen and release carbon dioxide. Photosynthesis and respiration are complementary processes. The organisms that originated these ingenious mechanisms were incorporated by symbiosis in the cells that protected them from the change in the atmosphere from a reducing, carbon dioxide-rich condition to an oxidizing, oxygen-rich condition. There is no evidence of the first life, except for these relics that are absolutely essential cellular organelles. The palaeozoic carbonates are the very last stage in this process, most of which happened long before.

Coal

Towards the end of the palaeozoic, a new way of sequestering atmospheric carbon occurred. The new land plants of the Carboniferous manufactured huge amounts of carbohydrates that were largely structural elements, not food, except for the bacteria who knew how to digest cellulose and live in the stomachs of plant-eaters. Cellulose is all energy-rich sugar, but locked up so devillishly that it is highly insoluble and very durable. This cellulose was buried in bogs away from the air that aerobic decay bacteria required for their functioning, and only those microorganisms that fermented remained, slowy abstracting the hydrogen and oxygen from the buried plant material, turning it into a colloidal jelly of what remained of the cellulose, which by now was something quite different. Deeper burial slowly squeezed out the water, and all bacterial action ceased, leaving only a complex black material of very complex structure, and consisting mainly of carbon. This material is coal.

Coal is produced only under special conditions of deposition, when the plant growth can flourish, followed by shallow swamps, followed by burial in sediments. The late Carboniferous (Pennsylvanian) was the first such time, when the conditions of sedimentation were cyclic, creating repetitive *cyclothems* of clay, coal and shale, in which the remains of the forests of *Lepidodendron* and *Sigillaria*, large fern-like plants, were preserved. Normally, aerobic decay processes would eliminate all trace of the former plants, returning their carbon to the atmosphere. The late Cretaceous was another such time. Most U.S. coals are either Pennsylvanian or Cretaceous. In other parts of the world, Triassic and Jurassic coals are found. Coal does not seem to be forming at the present time, although there are peat bogs, but no assurance that these bogs will eventually be coal. They could just dry up and disappear by decay.

A typical coal seam begins with the *seat earth* at the bottom, the original clay bottom of the swamp in which the forests grew. Then there is the coal, from numberless generations of strange trees, but no fossils of fish, showing that the swamp was black and lacked dissolved oxygen to support life. When trunks fell into the bog, they did not decay, but were attacked by the anaerobic bacteria that turned them first into a jelly, and then to peat, as the oxygen was used up and the hydrogen spent. By pressure, the peat was dried and by further bacterial action was altered to coal. Above the coal is fossiliferous sandy shale, or sandstone, showing that the bog had been submerged, the forest drowned and covered with sediment and fresh water. The upper part of the coal often contains evidence of the fallen trunks from the last generation. There may be a layer of iron ore, the result of more bacterial action when the forest was gone, and before freshwater conditions returned. This sequence is repeated over and over in the Coal Measures, as these rocks are known, and the seams extend over considerable areas.

Coal is a valuable fuel resource because it is elemental carbon, which can be burnt in the atmospheric oxygen to release heat. This, of course, only reverses the process that formed it, where the heat input was from the sun. Coal is just solar energy, released from where it has been stored for millions of years. There is only a finite amount of coal, and the sea and atmosphere can probably accommodate all of its carbon as carbon dioxide, if it were released at a reasonable rate. The problem is, however, is that humans tend to spend their money all at once, and burn all their coal the same way, which may cause a temporary disturbance. The carbon dioxide could be used to make more cellulose, again sequestering the carbon and releasing oxygen, while creating a new store of fuel. But humans will be long gone before this slow process can be realized.

Petroleum

A much smaller amount of carbon is sequestered as petroleum than as coal. Petroleum is to carbon as silica is to silicon. Silicon is not found in elemental form in nature, because it would much rather exist as silica, which has a much lower free energy. Carbon will combine with oxygen as well, to form the gas carbon dioxide and the carbonates derived from it. However, carbon will also combine with hydrogen to achieve a state of lower free energy than it can by forming graphite (and certainly diamond). Instead of the giant molecules of silica, we get the modest molecules of the hydrocarbons. The original source of the carbon must be living tissues, from which something must abstract the oxygen. When this is done, we have petroleum, which can be a gas, a mobile liquid, or a solid.

This seems to have happened when the black muds into which dead diatoms and other algae had fallen and had been preserved, began to change into shales. If this occurs in the absence of air, we have something similar to the conditions for the formation of coal, but the great masses of cellulose are missing, and we have the remains of bacteria, algae, and other simple forms of life. Microorganisms again go to work, removing the oxygen. As they do so, gases may be evolved, such as methane, CH_4. The same thing occurs when coal is formed, incidentally. In some coal beds, the methane has escaped, but in others its departure has been hindered, and the coal is called *gassy*. This gas is now being exploited in the United States where it exists in quantity. It will quickly be exhausted, but represents a windfall.

The liquids may escape from this shale, under the weight of overlying sediments, gas pressure, or floating on salt water, if there are permeable rocks adjoining. The carbon remaining turns the shales black, showing that they once held life, and that they were buried in anaerobic conditions. Most shales are not black, and whatever life was in them decayed and returned to the winds and waters before the rock was formed. In other cases, the black shales are clearly the *source rocks* for the liquid and gaseous petroleum that may be trapped nearby. Petroleum is much simpler structurally than coal. Of course, the original form of the carbon has no influence, since the petroleum has moved from the place it was formed. A liquid petroleum, *crude oil*, may be largely straight-chain hydrocarbons, or *paraffinic*, or contain 6-membered rings and double bonds, and be *asphaltic*. Underground, it may have smaller molecules dissolved in it that come out in bubbles when the pressure is reduced.

Petroleum is a double-barreled energy reserve, because both the carbon and the hydrogen can be burned to create heat. It is still mainly burning carbon, however. Natural gas, represented by CH_4, is 75% C and 25% H. Gasoline, represented by hexane, C_6H_{14}, is 84% C and 16% H. Liquid petroleum fuels are specially attractive because they are easy to handle and have a large energy density. It is much easier to carry around gasoline than the equivalent in storage batteries. Natural gas was easy to handle, but the energy density was so low that it was not economical to transport, and much was wasted, often flared off uselessly (but economically). The low price of fossil fuels makes such waste common. Using natural gas to fire boilers is a tragic waste of a valuable resource, but it is even encouraged. Petroleum is much more valuable as a chemical feedstock than as a fuel, which the world will discover to its sorrow in a century or so. Coal is the only fossil carbon whose fuel use can be justified.

Petroleum exists in several forms in the earth. Liquid and gaseous petroleum are the most known, because they are easily gathered. Petroleum-saturated tar sands exist in northwestern Canada. Huge amounts of gray-green shale are permeated by a waxy petroleum called *kerogen* in western Colorado and Utah. In fact, illuminating oil was first distilled from oil shales in Scotland, before liquid petroleum was widely known. The distinctive thing about tar sands and oil shales is that real *work* must be done to get at the petroleum, and this work will be expensive and environmentally damaging. We have broken into the bank of stored energy, and are only sweeping up the cash we find lying around in view, not sweating to pick locks and pry open drawers.

Liquid and gaseous petroleum, crude oil and natural gas, lurk under pressure in the pore space of sedimentary rocks. The rocks must not only be porous, but also *permeable*, so the petroleum can flow. This is usually the case, since the petroleum had to leak in in the first place. However, nature is satisfied if this happens in ten thousand years, and we are a bit less patient. If we drill a hole to a porous, permeable rock holding petroleum, it squeezes into the hole and usually up the hole to the ground. If we want gas, the gas must do this in any case. If crude oil only goes part way up, we can pump out what is in the hole and let the hole fill up again. You cannot *suck* petroleum out of the rocks. Of course, we can help. Gas can be forced down a hole to give extra pressure. The gas is usually carbon dioxide, which also makes the oil less viscous. Water can be forced in from wells surrounding the oil, to wash it forward. These are really desperation measures, but can extend the life of many oil fields that would have been abandoned in the past. If oil commanded a realistic price (perhaps twice or three times what it now brings) these efforts would be more general, and much oil would be saved that is now going to waste.

Exploration for oil is now a very refined and powerful process, and has rummaged literally everywhere, from pole to pole. However, there is no process that can detect the actual presence of petroleum except the drilling of a hole, and the nature and contents of the reserve cannot be estimated until some oil has actually been produced. It is always dangerous to make predictions of such uncertain things as the presence of oil, but saying that there are no more large, or even medium-sized oil fields left to be discovered in the United States is about on the same level as predicting that a large asteroid will not collide with the earth and wipe out a continent in the next century. Either could happen, but not bloody likely. The decline of domestic production, as a ten-year running average, is inevitable. Venezuela, Mexico, Canada and Nigeria will slowly follow, and eventually even Iran, Iraq and Saudi Arabia. The current demand for oil is huge and growing, the supply finite. It is like a pile of cash, so easy to scoop up as many armfuls as we want, but the pile just gets smaller. As late as 1973, the United States could dominate the world oil market, but now it is powerless. The United States was once extremely rich in cheap petroleum. Now we are scraping the sides of the bowl.

Charcoal, Coke and Carbon Black

Charcoal, coke and carbon black are all *colloidal* forms of carbon, and this should be kept in mind while considering them. Being colloidal means that they are effectively all surface, though this may not be evident in the bulk.

In Latin, *carbo, carbonis* was charcoal, and a *carbonarius* was a tradesman who made it. In German, *Kohl* was charcoal, and a *Kφhler* was a tradesman who made it. In English, *coal* was charcoal, and a *collier* a tradesman who made it. Charcoal, a smokeless fuel, was the preferred fuel for the home, used in a brazier to cook or to heat a room. Charcoal was also the fuel for the smelter of iron and other metals, and for the smiths who worked it at their forges, because it was pure. To make charcoal, hardwood trees were cut into sticks, which were piled in a dome-shaped heap and plastered over with mud and turf. A typical heap might include 15 to 45 cords of wood (a cord is 128 cuft). A hole was left at the top, and an opening at ground level on one side.

The fire was lit at the opening, and when things were

burning nicely, the opening was nearly blocked up to limit the air supplied to the fire. Dense smoke came out the hole in the top, as the volatiles were burnt to drive the rest out, at a temperature of about 400°C. It could take as long as a fortnight to complete the operation. When the smoke became light and blue, the job was done. The air was completely blocked off to extinguish the fire and the heap left to cool. When it was torn apart, much of the wood was now porous black sticks, mainly of carbon, with all volatile components driven off by the heat. The charcoal was nearly pure carbon, so it would burn without smoke or yellow flame, and not contaminate anything with impurities. The use of charcoal as a fuel is discussed under Fuels. A typical hardwood charcoal was 20 lb to the bushel.

While the pile was burning, a watery liquid would trickle out of the bottom. This was *pyroligneous acid*, 80%-90% water, but including formic and acetic acids, methyl alcohol, formaldehyde, acetone. Resin in the wood would give rise to turpentine and tar. The gas, of which some 8000 cuft were produced per cord of wood, was 59% CO_2, 33% CO, 3.5% CH_4, 3% H_2, and the remaining 1.5% heavier vapors. These were valuable by-products, but only feeble efforts were made to save and purify them. Now, charcoal is made in retorts and these products are recovered.

What we now call coal was used in local areas from ancient times, and was considered a kind of inferior fuel that gave odoriferous smoke. It was generally called *stone coal.* It was used in forges, for industrial purposes, in the 13th century in Liθge, an early centre of ironworking. Its widespread use began in the 17th century in England. Coal from the Tyne valley and the coast of Northumberland began to be exploited in bell pits, carried out by women with baskets, and was sent to London by sea. At this time, London had exhausted the forests in all directions, and charcoal was scarce. In fact, the use of charcoal for smelting had been outlawed, sending smelting to far-away districts, and even overseas to North America. This was called *sea-coal*, and there are references to it in Shakespeare.

The red-hot charcoal combines with the oxygen according to $2C + O_2 \rightarrow 2CO$, producing the colorless, odorless gas carbon monoxide. The carbon monoxide burns above the coal with a blue, almost invisible flame, according to $2CO + O_2 \rightarrow 2CO_2$. The fire is completely safe as long as it gets enough air, and that any little CO that happens to escape is removed by ventilation. Today's tightly-closed houses make indoor charcoal fires unadvisable, however.

The carbon monoxide is a very important feature in metallurgical fires. In the blacksmith's forge, it makes the atmosphere in the fire reducing, which means that iron heated in it will not oxidize or rust rapidly. In the iron furnace, it combines with the ore to give metallic iron. The iron ore is mixed with limestone, so the lime combines with the silica in the ore to make a slag that melts at low temperature.

When all the iron ore has been reduced to metallic iron by the carbon monoxide percolating through it, the iron and slag are removed as a spongy mass, which is then hammered to consolidate it and to drive out the slag. When it cools, it is reheated in the forge fire. This was the way iron was made from prehistoric times until the 19th century, by *direct reduction* in a charcoal fire. In the United States, it survived until the mid-19th century, until driven out by the more efficient blast furnace and puddling furnace. It was still used later in some places, such as Sweden, to make very pure iron, since the charcoal introduced no deleterious impurities.

Carbon dissolves in liquid iron. If the forge fire were blown too assiduously, and the sponge melted, the iron would absorb carbon from the charcoal, its melting point would drop, and the whole thing would form a molten ball. When this ball solidified, it was hard and could not be forged. Therefore, it was useless and the day's work was lost. In the 18th century, Abraham Darby II (1711-63) of Colebrookdale found that this disaster could be turned to advantage in making castings of the liquid iron, as people had long done with bronze, so it was not wasted after all. When the hot iron and carbon solution solidified, some of the carbon formed the white, hard substance Fe_3C, cementitite, and most of the rest made flakes of graphite in the iron. This *cast iron*, an alloy of iron and carbon, has remained of great importance. Abraham Darby III erected the first iron bridge at Ironbridge in 1779 from cast iron, and it still stands.

The first Abraham Darby (1677-1717) introduced coke into metallurgy, substituting an abundant resource, bituminous coal, for a scarce resource, charcoal, in 1709. Coke is to bituminous coal as charcoal is to wood, and is equally pure. This prevented contamination of the molten iron with sulphur and phosphorus, which had defeated earlier attempts to use coal for smelting. Darby's innovation, however, was not generally accepted for many years afterwards. Coke is discussed at more length below under Fuels.

Since ancient times, carbon had deliberately been put into iron by packing the iron in graphite, and placing it in a clay crucible carefully luted shut. When heated for an extended period at red heat, the iron absorbed a little of the carbon, a few tenths of a percent, but this was enough to change its properties drastically. When heated, this iron could be forged, like normal iron, but when heated and plunged into water, it became extremely hard. The new substance was steel (which we have seen gave its name to diamond). This is called *cemented* steel from the process of heating it with carbon. There were many different processes at different times and places, and the whole thing was largely a mystery surrounding trade secrets. It was, however, the key to effective weapons, as well as to ordinary tools like knives and scissors.

Fuels

The word "fuel" comes by a tortuous route from Latin *focus*, "hearth," via *focalia*, "things pertaining to a hearth." We burn the fuel to heat us and our food. Practically all of our fuels are carbon-based, for reasons that we will discuss in what follows. First, we have to look at how to express certain magnitudes associated with fuels.

A typical effect of burning a fuel is heating of water, and "heat" is expressed in terms of this. The *calorie* is the heat required to warm a gram of water by 1°C. The *Btu* (British thermal unit) is the heat required to warm a pound of water by 1°F. The relation between C and F degrees, and grams and pounds, make 1 Btu = 252 cal. Heat, we know, is actually energy, and measurement of the amount of energy required to heat water gives 1 cal = 4.186 J, and 1 Btu = 778.26 ft-lb. A watt is a joule per second, and a horsepower is 550 ft-lb per second, or 745.575 W. Heat cannot be transformed entirely into work under normal circumstances. The fraction that is converted to work in an engine is its *thermal efficiency*, η. The thermal efficiency of an automobile engine is around 25%, while an electric power station can have an efficiency of as much as 40%.

In engineering calculations, gases can be described closely enough by the ideal gas law, $pV = nRT$, where p is the pressure, V the volume, n the moles of gas in the volume V, T the absolute temperature, and R a universal constant whose value depends on the units chosen for the other quantities. Absolute temperature is expressed in kelvin, K, with K = C + 273.15, or in rankine, R, with R = F + 459.69. The values 273 and 460 are close enough for government work. Standard conditions are 0°C or 20°C and one atmosphere. If p is in pascal and V in cubic meters, then pV is in joule. R is then 8.314 J/gmol-K. A gmol is a mass in grams equal to the molecular weight. We need C = 12, O = 16, H=1, S = 32, N=14 and air=28.97. A gmol of any substance contains the same number of molecules, about 6.02×10^{23}, Avogadro's number. An *atmosphere* is 1.01325×10^{5} Pa, or 760 mmHg. If we use atmospheres to measure pressure, and liters (1000 cc) to measure volume, then R = 0.082 l-atm/gmol-K.

In the fps system, volume is measured in cubic feet (cuft), and pressure in pounds per square inch (psi), or atmospheres, where 1 atmosphere = 29.97 inHg. Standard conditions are often taken as 60°F and 30 inHg. The mol is now the pound mole, lbmol (which contains 453 times more molecules than the gmol). Convenient values for the gas constant are R = 0.7302 cuft-atm/lbmol-R or 10.73 cuft-psi/lbmol-R. A lbmol of gas occupies a volume $V = RT/p = 0.7302 \times 520 = 378.9$ cuft. From this, we can find the mass per cuft of any gas under standard condtions, if we know the molecular weight. Since the molecular weight of methane, CH_4 is 16, it weighs 0.04214 lb/cuft under standard conditions.

The *specific gravity* of a solid or liquid is the ratio of the density to the density of water, at specified temperatures. The specific gravity of a gas is often the ratio of the density to the density of air under the same conditions. For liquid fuels, the specific gravity is given in degrees API, where °API = 141.5/ spgr - 131.5. 10°API corresponds to spgr = 1.0, and the higher the number of degrees, the smaller the specific gravity. 40°API corresponds to spgr = 0.825.

To begin our consideration of fuels, let's first look at hydrogen, which is not a commonly-used fuel but is often suggested as one. Hydrogen burns in air according to $2H_2 + O_2 \rightarrow 2H_2O$, releasing 61,100 Btu/lb of heat if we condense the resulting water to a liquid, 51,623 Btu/lb if we do not. It takes 1050 Btu to vaporize a pound of water, and sometimes this energy is used, sometimes it is not. Every hydrogen-containing fuel has a high-heat value, where the water is condensed, and a low-heat value, where it is not. This is a nuisance in describing things, but it is easy to understand. A cubic foot of hydrogen under standard conditions contains 0.00264 lb, so burning it gets us about 161 Btu. A cubic foot of methane, the major constituent of natural gas, yields about 1000 Btu per cubic foot. Hydrogen power is not dense. It would be rather uneconomic to transport it by pipeline, or to store it in tanks.

Hydrogen forms an explosive mixture with air in the proportions by volume from 4% to 74.2%. The similar limits for gasoline vapour are 1.5% to 7.5%, for propane 2% to 10%, for methane 5% to 15%. This makes hydrogen an extreme explosion hazard. City gas used to contain hydrogen, and every winter city blocks would blow up here and there when the smallest leaks occurred. Hydrogen cannot be used as fuel for internal combustion engines, since it will not burn nicely—its antiknock properties are zero. It has been used as rocket fuel, however. It is expected to make a fuel cell fuel, but most small fuel cells burn things like alcohol, not hydrogen. I do not know if a reasonable hydrogen fuel cell has been devised. For all these reasons, hydrogen would not seem to have great promise as a general fuel.

There is another little drawback to hydrogen. There are no hydrogen mines. Most hydrogen on the planet is in the form of water—that is, it is already burned. To produce hydrogen, we have to put in more energy than we will receive by burning it. The economics of doing this with "free" solar energy are not good—the investment is large, and the hydrogen produced is small. In fact, the most economical ways to produce hydrogen at present use carbon fuels, by reforming hydrocarbons or by de-oxidizing water with hot carbon. Hydrogen at best is only an energy distribution system, not a prime energy source, and a rather inconvenient and dangerous system at that.

Solar energy is the source of most of the available energy on earth. It can be used directly to heat water, a use that is quite practical, if not responsive to variable demands. Sunlight can produce electricity with photovoltaic cells, which is also practical but expensive, and involves the inconveniences of storing electricity. Solar energy has been used for millennia, however, in the form of charcoal. Trees grow in the sun, then the wood is converted to charcoal, which is nearly pure carbon in a porous and easily used form. Charcoal is burned according to $C + O_2 \rightarrow CO_2$, making available about 14,000 Btu/lb. Energy can be stored in the form of wood, and used when necessary, not at the constant rate of direct solar energy.

Human population has already far outstripped charcoal resources. Forest competes with arable land, and is disappearing steadily, in place of growing or remaining constant. As early as the 16th century, wood shortages were felt in the vicinity of cities (then much smaller than today's). The recourse was to stored forms of solar energy that were laid up over millions of years, but could be exploited in centuries. These finite reserves have supported the economy that has now grown to huge dimensions, and is using the reserves at a rapid rate. These fuels, which I shall call carbon fuels to spread the net wide, contain mainly carbon and hydrogen, with smaller amounts of oxygen, sulphur and nitrogen. The carbon and hydrogen burn as described above, with a small deduction for the cost of separating them. Sulphur contributes only about 4000 Btu/lb, from the reaction $S + O_2 \rightarrow SO_2$. Fortunately, the amount of sulphur in carbon fuels is small, at most a few percent. Oxygen represents a part of the fuel that is already burnt, in effect, while nitrogen and what makes up the ash is inert.

The composition of a carbon fuel is given by a *proximate analysis* in terms of fixed carbon, volatiles, ash and moisture, or by an *ultimate analysis* of the dry fuel in terms of percentages of C, H, S, O, N and ash. The heating value of the fuel can be calculated in terms of the ultimate analysis. The Dulong equation is Btu/lb = 145.5C + 620(H - O/8) + 41S, while the Vondracek equation is Btu/lb = (160.5 - 0.112C')C + 486(H - O/10). The letters represent the percentages in the ultimate analysis, and C' is the percent carbon in a *solid* fuel. The heating value of a liquid petroleum fuel can be estimated by Btu/lb = 18,250 + 40 × °API.

Fuels are conveniently classified as solid, liquid and gaseous. Solid fuels are sold by weight, liquid and gaseous fuels by volume. We have already discussed charcoal. The other principal solid fuels are coal and coke. Coal is classified by *rank*. *Anthracite* contains more than 86% fixed carbon. It is hard, black and dustless. *Bituminous* either contains more than 69% fixed carbon, or has a fuel value of more than 11,000 Btu/lb. *Sub-bituminous* has a fuel value of 8300-13000 Btu/lb. *Lignite*, or brown coal, has a fuel value of less than 8300 Btu/lb. These ranks display less and less metamorphosis of the coal by heat and pressure. Sub-bituminous and lignite coals are subject to *weathering*, that is, disintegration on being exposed to the weather where they are subject to alternations of wet and dry. Bituminous and anthracite coals are not subject to weathering.

Many bituminous coals are called *agglomerating*, or *coking* coals because they swell upon heating, and then melt to produce a sintered cell structure, which forms a coherent *coke*. There is no necessary relationship between rank and agglomerating nature. Coals of lower rank, especially those that weather and contain pyrite, are subject to *spontaneous combustion* when stacked in a heap so that heat cannot escape. This is largely a function of the surface area presented for oxidation, so need not be feared for lump coal of high rank. Lump bituminous coal has an average density of about 50 lb/cuft, which includes the voids.

There are four main coal substances in bituminous coal. *Vitrain* is shiny and glassy, but very opaque. *Durain* is a dull, grey substance, hard and tough. *Clairain* is finely laminated vitrain and durain. The bright coal, vitrain and clairain, are separated by lenses of durain. These parts of the coal are clean and durable, but *fusain* is black and powdery, and responsible for all the dirt and dust. *Banded* bituminous coal, the common variety, consists of layers of vitrain and fusain. *Splint* coal is dull, consisting mainly of durain, and does not swell upon heating. The special coals *cannel* and *boghead* have unusually large amounts of volatile matter, but are noncoking. Cannel coal was made from pollen, boghead from seaweed. Boghead coal gives large amounts of tar and oil when distilled. Anthracite coal contains no fusain, so it is clean and burns with little smoke. However, anthracite is very difficult to light, while steam coal is much easier to get burning. The structure of coal is very complicated, and has not been well-deciphered. It is predominantly aromatic in structure, giving aromatic hydrocarbons upon distillation. The heating value of bituminous coal ranges from 10,500 to 14,400 Btu/lb.

Anthracite coal made an excellent stove fuel, better than charcoal, because it was denser and cleaner. It was the fuel of choice on the east coast of the U.S. until the coming of natural gas, and the exhaustion of the anthracite fields. It is difficult to light an anthracite fire unless you have experience with it. Kindling wood is required (liquid lighters will not suffice), and a significant amount of time. Once lighted, however, the fire would be steady and hot, with no smoke or fumes. Proper venting was required to avoid the carbon monoxide danger, as with charcoal. A good fire also requires equal-sized lumps. Mine-run coal was crushed (called breaking) and screened.

Bone was material from the mine with 40% to 75% fixed carbon, *slate* with less than 40%. Both were eliminated from stove fuel, which was carefully graded in uniform sizes. Particles smaller than 1/32", called *culm*, were not sold and were used for industrial purposes, such as locomotive and boiler fuel. Culm was a necessary consequence of breaking, and a large amount resulted. Culm is still being economically recovered from waste heaps. Anthracite was discovered in America near Pottstown, Pennsylvania in 1790 by hunter Necho Allen, but it was quite a few years before its proper use was understood, around 1808. The use of anthracite in iron smelting was first tried in 1795, but this was not perfected until considerably later.

An excellent fuel for metallurgical purposes is coke, made by driving off the volatile matter from bituminous coal. Coke can be made like charcoal, in *beehive* coke ovens, so-named because of their shape. The idea is to burn the volatile matter in the coal to drive out the volatile matter. When the smoke at the chimney turned from thick and brown to thin and blue, the coke was made. The oven was cooled and the coke scraped out. The gases given off were so valuable (and so noxious) that *by-product* ovens were made like coal gas retorts (see below), and the volatiles condensed and used.

Coking coal swells when heated as the volatiles are driven off, creating a strong, cellular structure. Coke is a dull dark gray to a light silver gray, much unlike coal, and is a kind of artificial anthracite, almost pure carbon. To be useful in the blast furnace, it must be strong to keep the charge permeable to gases. In Britain, railway locomotives were required for some years after 1830 to burn coke, so they would not emit smoke. Coke was a good, but rather expensive, fuel. A typical foundry coke might be 90.5% fixed carbon, 8.64% ash, with less than 1% left of the volatiles. Coke has a heating value of about 13,000 Btu/lb, being nearly pure carbon. The purification resulting from coking prevents impurities from entering the iron, especially sulphur and phosphorus.

Liquid petroleum, or *crude oil*, can be used as a fuel as it comes from the ground. Its value and utility are greatly enhanced, however, by *refining*. This is, essentially, fractional distillation. The crude oil is heated gradually in a still. The most volatile components come off first, and are condensed. The temperature is raised, and the less volatile components are condensed. This continues as the temperature is raised, until the still contains only a thick black tar, the "bottoms." Each component can be characterized by the temperature range over which it is collected. Present-day refining is not limited by the initial contents of the crude oil, since heavy components can be *cracked* into lighter components, and lighter components *reformed* into heavier.

Petroleum naphtha, or *gasoline* (a Standard Oil trade name) came off from 38°C to 200°C. It is about 85% C, 15% H, and has a heating value of 18,900 Btu/lb. Its specific gravity is 55° to 70° API (0.7587-0.7022), so it weighs about 6 lb per gallon. This *straight gasoline* has an ignition temperature of 350°C, and requires about 14-15 lb of air per lb of fuel to burn completely. It is about 78 octane. There was originally little use for this fraction, and it was burned around the refinery or flared. The worst antiknock behaviour is exhibited by straight-chain paraffinic hydrocarbons. The octane rating can be improved with branched chains, double bonds, or aromatic ingredients.

The next fraction, coming off between 160°C and 300°C, was called *kerosene* ("wax oil") by Standard Oil, and was the principal desired product, a fuel for lamps. To be safe in this use, the oil at normal temperatures should not have a high enough vapour pressure that a match would light the vapour before it reached the wick, possibly causing an explosion. A specification called the *flash point* was established to ensure safety. The flash point of a liquid fuel is the temperature below which the vapour will not catch fire under certain standard conditions. Kerosene must have a flash point above 73°F. In fact, it is usually from 110 to 130°F. Kerosene is a little heavier than gasoline, 40° to 48° API, about 6.7 lb per gallon. The heating value is the same as for gasoline, 18,900 Btu/lb. Kerosene is called "paraffin" in England, which was another trade name, this time of a firm that made it from shale oil in

Scotland. It is inevitable that some refiners would increase profits by putting as much naphtha in their kerosene as they could get away with.

The casualties from exploding lamps caught national attention, and provoked legislation. Standard Oil bragged that their kerosene was kept to a Standard, and this was actually more or less true, helping to lead the company to success (along with some other less creditable reasons). Gasoline has a flash point of -47°F, far below normal room temperatures, so gasoline is always ready to explode, which is necessary for its use in engines. This is an excellent reason for not using gasoline in lamps that are not specifically made for it, or for using gasoline as a dry-cleaning fluid, or for other household uses that may suggest themselves.

The heavy oils coming off after kerosene could be used as fuels or as lubricants. They made good lubricants, and were more pleasant than the animal oils and greases previously used. Fuel oils are now denoted by number. No. 1 fuel oil is little different from kerosene, while No. 6 is heavy and viscous, requiring preheating so it will flow. The flash points run from 100°F to 150°F. The pour point is the temperature at which the oil will flow more or less readily. These are the *distillate* fuel oils, the best of the sort. *residual* fuel oils are what is left in the still when the other fractions have been boiled off. They are likely to be black with suspended carbon, and solid at room temperature. *crude* fuel oils are crude oils unsuited to refining. Fuel oil has the advantage of being much easier to store and handle than coal, and is a very desirable fuel.

Crude oil and liquid fuels are sold in the United States by the barrel (bbl) of 42 gallons, about 5.6 cuft or 158.57 litres. This is the old wine barrel, and just happened to be the size of barrel that was around when the first drilled wells in northwestern Pennsylvania began production in the 1860's.

A liquid carbon fuel that does not come from petroleum is ethanol, C_2H_5OH, which can be made organically by the fermentation of sugar produced by the hydrolysis of starch, processes catalyzed by enzymes. Ethanol provides 12,770 or 11,585 Btu/lb, about 2/3 that of petroleum. As a motor fuel, it has the advantage of excellent antiknock properties. Special engine designs are not required if it is used as a gasoline supplement. A useful blend is 65% gasoline, 20% ethanol, and 15% benzene. The benzene is to prevent separation if water gets into the mixture. This blend has a heating value of 17,224 Btu/lb, about 91% of that for straight gasoline. Benzene itself, which can be derived from coal, is a good motor fuel, providing 17,480 Btu/lb (low) with good antiknock properties. In German, *Benzin* is the word for motor fuel.

A liquid fuel that is used as a gas is LPG, liquefied petroleum gas, often propane, C_3H_8. Its boiling point is -43.8°F, so it liquefies easily under reasonable pressure, and there is no trouble with vaporization when it is used. Its specific gravity is 0.508, so it weighs 4.23 lb per gallon. The vapour is 1.522 times heavier than air, and 36.5 cuft are obtained for each gallon of liquid. The heat of vaporization is 183 Btu/lb, and the heating value is 21,690 Btu/lb, or 2514 Btu/cuft, twice that of natural gas. Burners must be adjusted to allow for this difference of heating value. Propane requires 15.6 lb of air per lb of fuel. The explosive limits are roughly 2% to 8%, making it safer than natural gas. Propane and butane can be used in internal combustion engines with complete success. LPG is a very convenient fuel, worth its premium price.

Gaseous fuels are easily made from coal, coke or oil. The iron blast furnace is a copious source of gas. It is blown to burn the carbon in its charge of coke, iron ore and limestone, producing CO that reduces the iron to metal. The gas that issues from the top of the stack has a lot of nitrogen and carbon dioxide, but usually about 30% CO that gives it a heating value of 90-110 Btu/cuft. This *blast-furnace gas* is used to heat the air for blowing the furnace, to heat reverberatory furnaces, and for other purposes around the steel works. A mixture of steam and air can be blown continuously through a bed of glowing coke to make *producer* gas. A typical producer gas might be 27% CO, 12% H_2, 2.5% CO_2 and 55% N_2, with a heating value of 120-180 Btu/cuft. Producer gas is a clean fuel for heating pottery kilns, glass furnaces and similar facilities.

Blue gas or *water gas* is, however, gas that may be sold

for use outside the factory. It is made by blowing air and steam alternately through hot coke. The "blow" with dry air brings up the temperature of the coke to white heat. Then the "run" with steam takes advantage of the reaction $H_2O + C \rightarrow H_2 + CO$ to make a gas that is typically 50% H_2 and 40% CO, with 10% nitrogen and carbon dioxide, which are low because no air is used in the run phase. The heating value of this gas is 285-310 Btu/cuft. It burns with a blue flame, which gave it its name. An illuminating gas, one that burns with a luminous yellow flame, can be made by *carburetting* the water gas by adding the gases made by thermal cracking of a fuel oil. This was the gas normally supplied as city gas in later years. The high CO content made it possible to commit suicide by sticking one's head in the oven, but this no longer works, since natural gas is not poisonous. This is also the cheapest way to produce hydrogen gas for a hydrogen economy.

The original gas distributed for street lighting and home use from early in the 19th century was *coal gas*, made by distilling coal in horizontal cast-iron retorts. All of the volatile ingredients are driven off when the coal is heated to about 900°C, leaving a desirable *gas coke* in the retort. The condensable fraction of the gas driven off yields ammonia, benzene, coal tar, creosote and many other valuable carcinogenic substances. The gas made is about 49% H_2, 42% CH_4, 6% CO, and the usual 10% nitrogen and carbon dioxide. It provided 520-575 Btu/cuft. A very similar gas comes from by-product coke ovens, where conditions are optimized for making metallurgical coke instead of gas. The appropriate fraction of the liquid output was used as a lamp fuel under the name of *coal oil*, a name that was transferred to kerosene in later years.

Gas producers using oil could be made smaller and more convenient for portable use. They used oil and steam, and the water gas dual process, high temperatures, and the cracking of the oil to smaller fragments, to make excellent illuminating gas. One example was Pintsch gas, which had a heating value of 1300 Btu/cuft. Pintsch gas was widely used for lighting railway carriages.

Natural gas, which is largely methane, CH_4, gives 1100-1150 Btu/cuft when pure. Other light hydrocarbons are often present, and enhance the heating value of the gas. It burns with a blue flame, so is not an illuminating gas. What light there is comes mainly from C_2 molecules and the molecular fragments from the chain reaction. Most of the radiation is invisible. Methane is also called *marsh gas*, from its occurrence in marshes, were it burns as the Will-o'-th'-Wisp, and *fire damp* in coal mining, where it causes explosions. (Some say the Will o' th' Wisp is impure phosphine burning.) The discovery of the Welsbach mantle of cerium oxide around 1896 gave an excellent light, as good as that of an incandescent lamp, when heated with a nonluminous flame. Carburetion was removed from city gas when the mantle became common, and since then house gas has burned with a blue flame, like straight water gas. Natural gas, of course, does not need to be manufactured. It comes out of the ground when you punch a hole in the right place. Because natural gases from different sources have different amounts of carbon dioxide and so different heating values, gas is sold by heating value, not by volume. A *therm* is 100,000 Btu, or about the heating value of 100 cuft, called CCF in the trade. An MCF is 1000 cuft. The volumes are figured at standard conditions.

Another curious gas is acetylene, C_2H_2. It can be made as required by dropping H_2O on CaC_2, calcium carbide. Acetylene burns with a bright yellow luminous flame. Acetylene lamps using carbide were once common as bicycle lights and helmet lights for miners. They included a steel wheel and "flint" like a cigarette lighter, so all you needed was a can of carbide and a little water. They were so excellent that I am surprised they are no longer common. Carbide is formed by heating lime and coke in an electric arc furnace: $CaO + 3C \rightarrow CaC_2 + CO$. Acetylene is formed in the reaction $CaC_2 + 2H_2O \rightarrow C_2H_2 + Ca(OH)_2$. The slaked lime did make a bit of a mess if you weren't careful. The explosive limits of acetylene are 2.5% to 80%, so it is as bad as hydrogen in this respect. Acetylene burns with a hot blue flame when pre-mixed with oxygen. The oxyacetylene torch is used for welding and other purposes, and the acetylene is supplied in tanks. It is dissolved in acetone, since if you just tried to pressureize the tanks, the acetylene would blow up.

BUCKYBALLS AND NANOTUBES

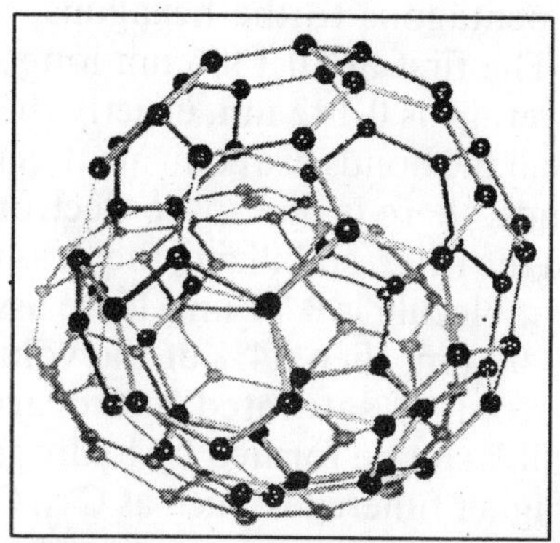

Buckminsterfullerene is the carbon molecule C_{60}. It is a convex polyhedron with 12 equal pentagonal faces and 20 equal hexagonal faces, each pentagon surrounded by 5 hexagons.

A model is shown at the right. Its symmetry is icosahedral, point group I_h, with 12 five-fold axes, 20 3-fold axes, 15 2-fold axes and 15 planes of symmetry, of order 120. This group has irreducible representations of dimension 5, 4, 3 and 1, which will be the degeneracy of states of the corresponding symmetries. It has the same symmetry as the common football (soccer ball), and is nicknamed the "buckyball." The fascinating story of its discovery can be read in Baggott. A plane can be tessellated with regular hexagons, but a sphere requires the addition of pentagons, as was shown by Euler. He showed that $F + V = E + 2$, where F = number of faces, V = number of vertices and E = number of edges. For buckminsterfullerene, $V = 60$, $F = 32$ and $E = 90$.

Buckminsterfullerene is named from its resemblance to the "geodesic domes" designed by R. Buckminster Fuller (1895-1983), of which a famous example was the U.S. Pavilion at the Montreal World's Fair in 1967. He began to design domes in the 1920's, but not until after the 1947 patent did he concentrate on them. Over 2000 have been built. The domes are assembled from tetrahedrons and octahedrons of lightweight material, and generally have hexagonal faces, but pentagonal faces are there.

There are two classes of carbon-carbon bonds in C_{60}, 30 connecting the pentagons to the hexagons, and 60 forming the pentagons. The first are 0.1386 nm long, the rest 0.1434 nm long. The average is 0.142 nm, exactly the bond length in graphite, where all the bonds are equivalent. Since each carbon forms three bonds, there is one spare electron per carbon to resonate around the rings, stabilizing them, as in graphite. The diameter of the molecule is 0.71 nm. If we assume the radius of a carbon is 0.07 nm, then 54% of the volume (about 0.19 nm^3) of C_{60} is void. A saturated hydrocarbon, $C_{60}H_{60}$, a "fuzzy buckyball," can be formed by hydrogenation.

There are higher fullerenes, such as C_{70}, C_{76}, C_{78} and C_{84}, composed of hexagons and pentagons with various symmetries. Fullerenes have been made inside fullerenes, so that they are concentric. Nested fullerenes can be made that resemble onions 50 nm in diameter, with 70 shells. In the work of synthesizing these fullerenes, needles were noticed that were found to be composed of *nanotubes*. These are helical assemblies of graphite hexagons. If the pitch is just right, the hexagons fit together smoothly (you might investigate this by rolling up paper with a hexagon net drawn on it). These were first seen in 1991, but similar structures had been noticed as early as 1960 by Roger Bacon at Union Carbide, also produced in an electric arc with carbon electrodes, when studying the much larger carbon whiskers that have found use in composite materials. Ijima, the 1991 discoverer, was the first to recognize that they were closed, smooth tubes, not rolled-up graphite.

The ends of the tubes are capped by spherical surfaces containing the necessary 12 pentagons, 6 at each end, while the tubes themselves are made only of hexagons. The tubes may be multiple, one inside the other. They are only a few nm in diameter (2-20 nm is usual), from which they acquired their name. They are, as may be expected, extremely strong—

stronger than carbon fibres or any other known material, for their size. Their growth apparently starts from a pentagon in one cap as one of the helical (chiral) fullerene structures, and then finds it easier to continue as tube of hexagons when the equator is reached. Buckyballs, buckytubes and buckyonions show that graphite would rather not be flat when it is in small amounts. All of the structures avoid the energetically unfavorable dangling bonds at the edges of a sheet, because they have no edges.

Buckyballs and nanotubes are much more interesting than they are useful, but ingenious uses may yet be found. Buckyballs can clathrate (jail) things, and nanotubes can conduct electricity and contain metal fillings. These properties may lead to applications. They were detected in the interstellar medium, produced by red giant carbon stars, which send soot into space. In fact, the study of interstellar absorption led to the discovery of buckyballs. These interstellar buckyballs seem to clathrate lanthanum.

Radiolarians are microscopic animals of the plankton that live in silica geodesic domes composed of hexagons and pentagons, and the occasional square, which greatly resemble fullerenes. *Reticulum plasmatique* and *Aulonia hexagona* are illustrated in Baggott. The pseudopoda extend from the openings to feed. A hexagonal net can tessellate a bathroom floor, but if it is to form a closed surface, pentagons must be introduced, as Euler proved.

Chapter 6

Silicon

Silicon is an word known to the public, at least in the phrase "Silicon Valley," and is associated with computers, although the public probably does not know what a chemical element is, or how silicon is used in computers, or where it comes from, or how it behaves. The earth is constructed of silicon, and we use it every day in the form of glass and pottery. What carbon is to the living world, silicon is to the non-living. Its study is, therefore, interesting and wide-ranging, leading into many areas of technology and science. In this article, I shall strive to present the fundamentals of the most important of these areas, so that a useful understanding of silicon can be developed in those who seek it. Not only is there a great amount of information about silicon, but study brings up new questions at every turn. I am certain that somebody knows the answers to most of these questions, but I do not, and must leave them for further research. Indeed, the authorities often do not agree, figures may vary, and ignorance is glossed over in many cases. I shall do the best I can, and admit my ignorance where it appears.

The physical and chemical properties of silicon and its oxide silica. If it were not for the semiconductor applications of pure silicon, silica would be by far the more important topic. The earth is more precisely constructed of silica; except possibly for the core, the whole earth is mainly silica, with a few impurities. Then, the use of elemental silica as a semiconductor will be discussed, and how several familiar semiconductor devices made from it work. After this, the large field of silica minerals and their structures will be explained.

Then, the relation of silica to life will be explored, concentrating on the diatom and the diatomite it has left, which is an important industrial mineral. Silicate minerals weather to clay, which not only forms the soils in which plants grow, but also gives us pottery, bricks and cement. Silicon also gives us glass, a substance that is probably always in our view in daily life, and which is not only useful, but also beautiful. The lore of silica is a rich garden that yields many tasty fruits.

The name *silicon* comes from the Latin *silex, silicis*, the word for "paving stone." The genitive is given to show where the "c" comes from. Flints made a very good, durable paving stone, so the word came to be associated with flints. Flints are made of silicon dioxide, so the name was acquired by this important substance, "silica." Silica is very difficult to decompose into silicon and oxygen, so was long supposed to be a simple substance, an "element," along with similar compounds, such as alumina, Al_2O_3, or magnesia, MgO. As sand, it has been known from time immemorial. The Greek for sand, *psammites*, and the Latin, *arena*, have not found their ways into chemical terminology as the flint has.

When Humphrey Davy decomposed soda and potash into metallic sodium and potassium by electrolysis just after 1800, powerful reducing agents became available. A reducing agent is a reagent that supplies electrons, which can turn a metal ion into the metal itself, as $Fe^{++} + 2e^- \rightarrow Fe$.

Potassium is just about the most powerful reducing agent that exists, and can be used to reduce nearly every other element. Davy, and in Paris Gay-Lussac and Thınard, competed to decompose previously undecomposable substances and discover new elements. Davy won with boron in 1809, closely followed by the French pair. Though Davy suspected silica was a compound, Gay-Lussac and Thınard won the race for silicon in 1811. What they got was not all that impressive, and Berzelius, in 1824, finally exhibited relatively pure silicon, and is credited with the discovery.

What these workers did was dissolve silica in hydrofluoric acid (the only acid that attacks it) to get silicon fluoride: $SiO_2 + 4HF \rightarrow SiF_4 + 2H_2O$. The gaseous silicon fluoride was then passed through molten potassium, where the silicon was reduced: $SiF_4 + 4K \rightarrow Si + 4KF$. The potassium fluoride could then be washed away, leaving a brown powder, *amorphous silicon*. It is not actually amorphous, just microcrystalline. The actual reactions are not this neat, and other compounds are formed, such as fluosilicates, which means that the silicon is not very pure. It was not until 1854 that pure, crystalline silicon was produced. This is a silvery-gray, metallic-looking substance that is, nevertheless, rather brittle and low in density.

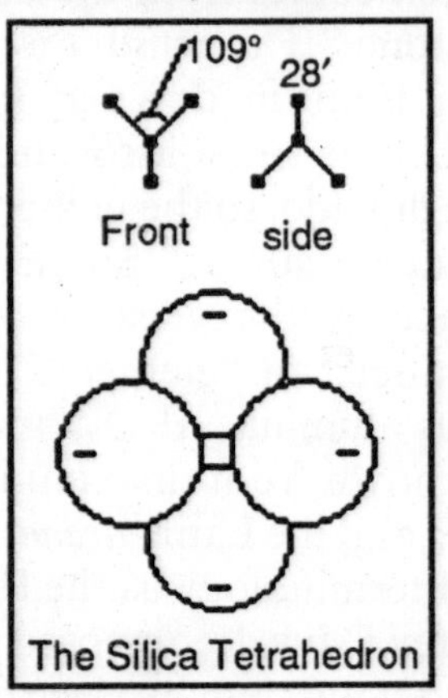

The Silica Tetrahedron

Most silicon now is made by reduction of SiO_2 with C in the electric furnace. With carefully selected pure sand, the result is commercial brown silicon of 97% purity or better. This is the silicon used for semiconductors, but it must be further purified to bring impurities below the parts-per-billion level. Both sand and the brown silicon are the starting points for the synthesis of silicon compounds.

Silicon belongs to the IVA family in the periodic table, which consists of carbon, silicon, germanium, tin and lead. Its atomic number is 14, and its atomic weight is 28.086. Its isotope Si^{28} has an abundance of 92%. It melts at 1440°C, and boils at 2355°C. Its density is 2.36 g/cc. The outer electron configuration of each of these atoms is s^2p^2. Carbon and silicon are nonmetals, white tin and lead are metals, and germanium and gray tin are typical *semiconductors*. Of course, the important use of elemental silicon is as a semiconductor, so this matter will be discussed in its own section below. These five atoms are very different in their physical and chemical behaviors, despite being in the same column in the periodic table.

Silicon is often compared to carbon, and differences are what is significant, not similarities. Each has a valence of 4, but there the similarity ends. Carbon will share one, two or three electrons with another carbon, forming single, double and triple covalent bonds. The lengths of these bonds are typically 0.154, 0.133 and 0.120 nm. The radius of a carbon atom can be taken as 0.077, based on the single bond. Multiple bonds are also formed with other atoms, such as N, O, S and P. This leads to the great richness of carbon chemistry. Carbon will not let go of its four valence electrons, but is willing to share them.

Silicon, on the other hand, will share an electron with another silicon, showing a radius of 0.117 nm, but will not make multiple bonds. The fact that it is about 50% larger than a carbon atom makes the difference. Silicon, likewise, will not let go of its four valence electrons, and does not even like to share them very much if the alternative of sharing them with an oxygen atom is possible. Both carbon and silicon find that hybrid sp^3 *tetrahedral* orbitals are the most stable, with the angle between bonds equal to 109° 28', the *tetrahedral angle*. Silicon prefers above all else to surround itself with four oxygen atoms as if in the *orthosilicate* ion $(SiO_4)^{----}$. This ion actually exists in water solution, forming the weakly ionized orthosilicic acid, H_4SiO_4, with ionization constant $K_1 = 2 \times 10^{-10}$. The orthosilicate ion, which we shall call the *silica tetrahedron*, is shown in the diagram at the right. All you can see of it are the oxygen atoms; the silicon is safe in the middle.

Molecules of orthosilicic acid can *condense* at their vertices by the elimination of H_2O to form Si-O-Si linkages. Each silcon atom is surrounded by four oxygens, in each of which it has a half-interest, so the composition is expressed by SiO_2, the usual formula given for silica. Note carefully that there are no SiO_2 molecules here, just a giant macromolecule that is the crystalline silica. If not given a lot of time, the tetrahedra will condense in a sort of mess that is not actually crystalline, but a *glass*. With time and luck, or if allowed to grow on a pattern, the tetrahedra will arrange themselves in the form of a crystal, perhaps the hexagonal crystal of the mineral quartz. The density of quartz is 2.65 g/cc. Silicon specializes in these large macromolecules, of which the silica tetrahedron is the building block.

The analogous compounds CO_2 and SiO_2 are very different. The first is a gas, or a soft solid at low temperatures, always consisting of CO_2 molecules, in which the carbon and oxygen are connected by double bonds. It is soluble in water, forming carbonate ions. The second is a hard, crystalline solid not softened by heat below 1700°C, and very resistant to chemical attack. Silica is one large molecule, not an assembly of SiO_2 molecules. I have not heard exactly what happens in the gaseous state.

If there is no oxygen around, the silicon atom must be satisfied by others of its own kind. The Si-Si covalent bond is 0.117 nm long, and is strong. The crystal formed is the same as in diamond, a face-centered cubic lattice. Its density is 2.36 g/cc. Large crystals are not as reactive as the brown powder of amorphous silicon (which is also fcc, but finely divided). The crystal is silvery gray, and shiny like a metal, but is lighter that you would expect from its appearance. One handbook mentions a "graphitic" silicon, but all other sources I have consulted are silent on this subject. If it were not for the electrons, pure crystalline silica would be transparent.

Silicon can be convinced to combine with halogens and hydrogen if there is no oxygen around. This combination must usually be brought about indirectly, and water must be excluded. Silicon forms *silane*, SiH_4, which is analogous to *methane*, CH_4. However, the two gases are quite different. Methane can be bubbled through water, but silane is immediately hydrolyzed: $SiH_4 + 3H_2O \rightarrow H_2SiO_3 + 3H_2$, since silicon has a hunger for oxygen. There are analogues to the alkanes, such as Si_2H_6, Si_3H_8, and so on, where there are Si-Si bonds. The last-mentioned compound, and all higher ones, are *pyrophoric*, bursting into flame even with only atmospheric oxygen available. Treatment of brown silicon with chlorine makes $SiCl_4$.

This substance melts at -70°C, and boils at 60°C. It is tetrahedral, with bond length 0.201 nm, slightly less than expected. The contraction of about 0.016 nm in the bond length is evidence of additional stability conferred by *resonance* with slightly ionic structures in which electrons are moved from chlorine to chlorine. In water, $SiCl_4 + 3H_2O \rightarrow H_2SiO_3 + 4HCl$. If $SiCl_4$ is mixed with NH_4OH, we get a dense white smoke of the metasilicic acid and ammonium chloride, that is used for military purposes and in skywriting. $SiCl_4$ liquid is nonconducting, showing that it is not ionic.

SiF_4 is curious in that its melting point is -90°C, but its "boiling point" is -95°C. What this means is that the solid sublimes to SiF_4 molecules before it melts. The solid does not consist of SiF_4 molecules, but of a giant molecule in silicon style. SiF_4 is covalently bonded, and does not give fluoride ions in solution—instead, the molecule hydrolyzes, as we would expect, making metasilicic acid and fluosilicic acid (instead of hydrofluoric). Fluosilicic acid, H_2SiF_6, is a strong acid where the silicon is in the centre of an octahedron of fluorines, instead of a tetrahedron of oxygens. The silicofluoride ions are stable in solution, and form salts. The silicofluoride octahedron is the only thing that can be on equal terms with the silica tetrahedron. SiF_4 is evolved when silica dissolves in HF.

There are also numerous compounds with both hydrogen and halogens, such as trichlorosilane, $SiHCl_3$. This compound is made when brown silicon is treated with anydrous HCl. It decomposes on a hot surface to elemental silicon, chlorine and hydrochloric acid. In the purification of silicon, it can be distilled to high purity, and then decomposed to silicon with impurities at the parts-per-billion level, satisfactory for many purposes.

Sand will also react directly with carbon in the electric furnace to produce silicon carbide, SiC, or *carborundum*: $SiO_2 + 3C \rightarrow SiC + 2CO$. This reaction proceeds because the gaseous carbon monoxide expelled. Commercial carborundum is black and impure, but still is very hard, 9.2 Mohs. When pure, it is a transparent, greenish crystal. It occurs in two modifications, α-SiC which is hexagonal in structure, and β-SiC, which is face-centered cubic, the diamond structure, and apparently the hardest form. Its electrical conductivity is 107-200 Ω-cm, so it can be used for electrodes. Its density is 3.217 g/cc, and it sublimes at 2700°C. Carborundum was discovered by E. G. Acheson in 1891, and was the first of the artificial abrasives.

Ferrosilicon, an alloy of iron and silicon, is produced if iron oxide is added to the sand and coke in the electric furnace. This alloy does not have to be especally pure, since it is thrown into molten iron to deoxidize it. The silicon combines with any oxygen present, as we might expect. Extra silicon remains and alloys with the iron. A few percent gives an excellent steel for magnetic cores, and in cast iron it aids the castability.

Sodium oxide, Na_2O, reacts rapidly with water to make the strong base NaOH: $Na_2O + H_2O \rightarrow 2NaOH$. NaOH dissociates completely in water to give OH^- ion, and because of this is called a *base*. Sulphur trioxide, SO_3, also reacts rapidly with water to make the strong acid H_2SO_4: $SO_3 + H_2O \rightarrow H_2SO_4$. This molecule dissociates in water to give H^+, and because of this is called an *acid*. Chemistry texts often point out that this ion is hydrated. This is no surprise: all ions in water are hydrated. If we mix the two, we get sodium sulphate and water, which is neither acidic nor basic. Therefore, we call sodium basic and sulphur acidic, representing two principles that will react with each other.

Calcium oxide, lime, CaO, gives $Ca(OH)_2$ in water. Incidentally, this is called *slaking* the lime and releases much heat. Calcium hydroxide is a weak base, but we have no trouble classifying calcium as basic. Silicon oxide, SiO_2, does not actually react with water, but we can imagine that it would form H_2SiO_3, a feeble acid, and so classify silica as acid. When the two meet in a fused state, they react to form Ca_2SiO_3, which has a lower melting point than either reagent. In general, acidic and basic oxides react to give a *slag* that is usually of relatively low melting point. This reaction is used to remove silicates in the smelting of iron. Limestone is added, which burns to lime, evolving carbon dioxide, and then the lime reacts with the silica to form a slag in which the iron is insoluble, and which can be drawn off separately. The use of the terms acid and base in this case, where there is no water to be seen, can be made reasonable by the preceding argument.

Orthosilicic acid can lose water to form metasilicic acid, H_2SiO_3, which we have encountered several times already. Silica is attacked by alkalis, $SiO_2 + 2NaOH \rightarrow Na_2SiO_3 + H_2O$. If we now treat this with HCl, we get $Na_2SiO_3 + 2Hcl \rightarrow H_2SiO_3 + 2NaCl$. The metasilicic acid is here called "water glass," since it forms a gelatinous precipitate that is insoluble in water. Water glass is actually formed from chains of silica tetrahedra in an irregular and indefinite pattern. The formula gives only the composition, not the molecule. This is, as we realise, typical of silicon. This gel can be used to seal the pores in eggs to preserve them longer, or as an adhesive. If heated, water is driven out of the abundant pore space, and the result is the familiar *silica gel* used as a desiccant. When it gets full of water, it can be renewed by simple heating. Sodium silicate is made commercially by fusing sodium carbonate, sand and carbon. Water glass is then produced by acidifying the product.

Silicon makes giant molecules, as we have seen. To get some perspective on atomic sizes, consider a picogram (a millionth of a microgram) of silica. This would fit in a cube 722 nm on a side, so it would be at the limit of the optical microscope if you wanted to see it, and almost as small as a colloidal particle. However, this speck would still contain 10^{10} SiO_2 units, 10,000 million of them! Silica vapour could easily consist of clumps of SiO_2 units, say hundreds of them in each fragment, but I have not heard of anyone who has investigated. The boiling point of 2230-2590°C makes measurements a little difficult. Crystalline silicon has 5×10^{22} atoms per cc.

There is another oxide of silicon, silicon monoxide, SiO. It forms cubic crystals of density 2.13 g/cc that are transparent, with index of refraction 2.0. Polycrystalline SiO is used as an optical thin film, usually to improve reflectivity. It melts somewhere above 1700°C, and boils at 1880°C. I have seen little information on its structure and technology. Apparently it can autooxidize according to $2SiO \rightarrow Si + SiO_2$.

A *silicone* is a silicon-oxygen chain with hydrocarbon radicals attached to the silicons, as in methylsilicone, $(CH_3)_3Si-O-(CH_3)_2Si-O-...-O-Si(CH_3)_3$. The hydrocarbons make the molecule look like a hydrocarbon to its surroundings, while the strong silicon-oxygen chain makes it very stable at high temperatures. The viscosity of a silicone oil increases much less rapidly than that of a hydrocarbon oil when the temperature decreases, say a factor of only 70 from 100°C to -35°C, while a hydrocarbon oil's would increase by a factor of 1800. The chains can be cross-linked and polymerized by oxygen or other means to give rubbery solids that are equally inert. These remarkable compounds are strictly artificial, never found in nature.

Silicon and its compounds are not poisonous or otherwise hazardous, except for rare compounds like the silanes. Most, indeed, are almost completely insoluble. One exception is the industrial disease *silicosis*, which affects workers exposed to silica dust over a considerable period. The workers involved are mainly stoneworkers and miners, and the disease can be prevented by wearing respiratory filters. Some years ago, a Denver TV reporter found to her horror that children's playgrounds often contained deadly silica, and that nobody was doing anything about it.

SILICON AND ELECTRONICS

For electronic applications, we are interested in elemental silicon as a conductor of electricity. By adding controlled amounts of impurities we can determine the sign of the charge carriers and their density. By doing this in well-defined limited areas, we can construct electronic devices that rectify, amplify, are sensitive to light, emit light, and store and move electric charge on command.

The silicon must be a single crystal with few defects, and of extreme purity, with impurities at the sub-parts-per-billion level. To produce this semiconductor silicon, we start with brown commercial silicon of about 97% purity. The silicon is treated with anhydrous hydrogen chloride to produce a chlorinated silane, which is fractionally distilled to high purity, and decomposed on a hot wire. The silicon is scraped off and cast into a polycrystalline rod. This relatively pure silicon is then suspended vertically and a molten zone is passed repeatedly from top to bottom. The heat comes from induction of a current by a radio-frequency current in a movable coil surrounding the sample. Impurities tend to prefer the molten zone to the silicon solidifying behind it. At the finish of the refining, the end of the rod containing the impurities is sawed off. This zone refining produces ultra-pure silicon. A seed crystal is placed at the lower end of the polycrystalline bar, and zone melting allows the growth of a single crystal based on the orientation of the seed crystal. In all of this processing, the silicon has never touched a crucible wall or the equivalent, and remains pure. The cylindrical crystal is then cut into thin slices, or *wafers*, for the fabrication process that follows.

Other materials have been used for semiconductor devices. The first transistors and diodes were made from germanium. Gallium arsenide, GaAs, an example of a *compound semiconductor*, has been used for specialized devices. Similar compound semiconductors must be used for optoelectronic devices, such as light-emitting-diodes, for which silicon is not suitable. Silicon, however, has remained by far the most widely used semiconductor because of its unique advantages. Among these are (1) it is elemental, so composition is not a problem; (2) it forms a tough, adhering insulating oxide, SiO_2, that can be used for isolation and protection; and (3) a large variety of impurity atoms are available for controlling its conduction properties.

The charge carriers in a metal that is a good conductor are electrons, which seem to move almost as if they were free particles, but confined to the volume of the metal by a high potential barrier on the order of 8 or 10 eV. At absolute zero, when the metal ions are in their regular lattice positions, the ions have no effect on the electron motion, except perhaps for changing their effective mass. In any real metal, there are impurities and defects with which the electrons may collide, and above absolute zero the vibrating ions also hinder the electrons.

These realities cause electrical *resistance*. The number of states which the electrons may occupy is limited, and there are usually so many electrons that all the states are filled up to an energy called the *Fermi level*, which is about 5 eV. This effect is called *degeneracy* (not a very good term, but it is the one used).

In any crystal lattice, metal or otherwise, electrons free to move about occupy states classified by their energy. The periodicity of the crystal lattice causes gaps in the allowed energies, separating them into *bands*.

This only applies to the electrons that can roam throughout the crystal, the valence electrons.

The electrons in the inner shells of the ions stay where they are and have no other influence. The band in which the valence electrons move is called the *valence band*. In a metal, this band is mainly empty. The electrons may be degenerate at ordinary temperatures, but they still move like free electrons.

In a silicon crystal, the valence electrons are the four outer electrons that form the covalent bonds between the silicon atoms. The possible states in the valence band are completely full.

There are lots of electrons, but they can go nowhere. As soon as they would move, another electron would take their place, and the net effect would be zero. In this state, the silicon is an *insulator*. This is almost true in pure silicon at low temperatures. The next higher band of electron states is called the *conduction band*. It is, in this state, completely empty. The distance in energy between the top of the valence band and the bottom of the conduction band is, in silicon, 1.12 eV.

As the temperature rises, electrons are jostled about by the vibrating ions. At room temperature, the jostling only amounts to 0.025 eV, so it is not very likely that an electron will be knocked upstairs. This is one of the advantages of silicon, since it would require a very high temperature before many electrons were jolted into the conduction band by thermal agitation. Once this happens, we lose control of the conduction characteristics by adding impurities, and the semiconductor becomes useless. In germanium, the energy gap is only 0.67 eV, so germanium gave up at much lower temperatures than silicon does. This is an important reason why silicon has almost completely replaced germanium in semiconductors. Diamond has a band gap of 5.4 eV, so it would be useful to high temperatures, and make excellent devices. Unfortunately, it is very hard to grow perfect diamond crystals, and difficult to find impurities that will control its conduction properties.

The conductivity produced by charge carriers thermally excited across the energy gap in a semiconductor is called *intrinsic*. Let's suppose for a moment that we have n_i electrons in the conduction band, and an equal number of places in the valence band where they used to be, called *holes*. If we apply an electric field, the electrons move one way, and the holes move the other way, both contributing to the current in the direction of the field. So far as semiconductors are concerned, holes act like positive charges with positive mass, very much like particles.

There is always a chance that an electron in the conduction band will encounter a hole in the valence band and fill it, or recombine. Two intrinsic charge carriers are lost in this process. The rate of recombination is proportional to the product of the concentrations of electrons n and holes p: charge pairs lost per second per unit volume = npr(T), where r(T) is some function of the temperature T. This is balanced by thermal generation of pairs at the rate g(T), which does not depend on the concentrations. In equilibrium, then, npr(T) = g(T), or np = g(T)/r(T) = f(T). In our intrinsic semiconductor, $n = p = n_i$. f(T) does not depend on the concentration, and is a constant in a particular crystal, so we have in all cases $np = n_i^2$. For silicon at 300K, $n_i = 1.4 \times 10^{10}$ per cc. Extreme purity of the semiconductor is necessary to achieve a figure this low.

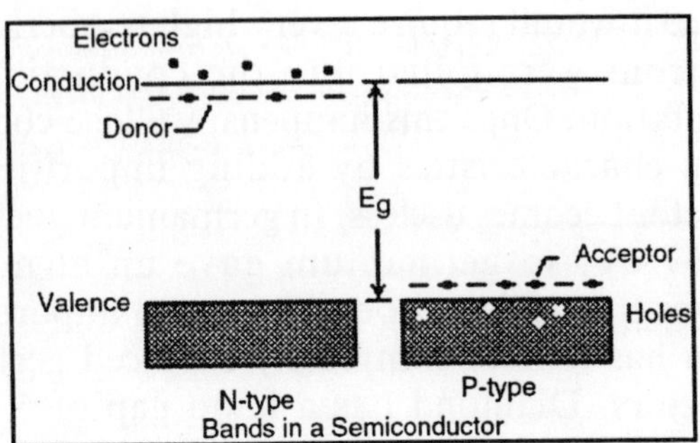

Bands in a Semiconductor

Now suppose we introduce a substitutional impurity in the silicon lattice that has one more valence electron than silicon. Examples are P or As, either of which will fit pretty well. Impurities that are deliberately added to control the characteristics of the semiconductor are called *dopants*. P or As can be added in amounts up to 10^{21} per cc before coming out of solution, which is many more than we will need.

Four of the valence electrons do the job of making the tetrahedral bonds to the other silicon atoms, while the fifth is left idling. It happens that it only requires about 0.4 eV to kick the fifth electron away into the conduction band, and this is easily done by thermal agitation at room temperature. This gives us a way of putting exactly the number of electrons we want into the conduction band. Accordingly, the impurity is called a *donor*. The diagram shows the filled valence band, the empty conduction band, and the donor levels, five of which are *ionized* and have given electrons to the conduction band. At the same time, from the equilibrium equation, $p = n_i^2/n$ will be very small. The *majority carriers*, electrons in this case, will far outnumber the *minority carriers*, the holes in the valence band, which can be neglected in a first approximation. Silicon with a pentavalent dopant is called N-type silicon.

On the other hand, we may use a dopant that has only three valence electrons, such as B, Ga or Al. This causes a lack of one electron in one of the covalent bonds made by the dopant. It requires only about 0.4 eV to jolt an electron into the deficiency, leaving a hole in the valence band to wander around. We can add up to 5×10^{20} boron atoms in the crystal without causing a lattice disturbance, which again is sufficient for our purposes. As many holes as we want can be introduced as majority carriers, while minority carriers, the electrons, fall to a low concentration $n = n_i^2/p$. These dopants are called *acceptors* because they accept an electron to create a mobile hole. These levels are only just above the valence band in the figure, and have accepted four electrons. The holes are shown in the valence band. Silicon with a trivalent impurity is called P-type silicon.

There are many ways to construct semiconductor devices. I shall describe the widely-used *planar* process here, in its original form. It has been greatly developed and refined to produce smaller and smaller devices, but the principles are the same. The surface of the wafer can be *oxidized* to create a tightly-adhering layer of SiO_2 by heating it in an atmosphere of oxygen. This layer is called simply *oxide*. Such a layer *passivates* or protects the silicon, and also serves as a mask for the application of dopants. Suppose we want to make a certain area of the wafer P-type by doping with boron. The oxide is coated with a *photoresist*, which is exposed to ultraviolet light through a photographically prepared mask. The area we want to dope is protected from the UV, while the rest is polymerized and hardened. Now the unpolymerized resist is washed off, leaving our region uncovered. The exposed oxide is etched away with dilute hydrofluoric acid, leaving the bare silicon exposed. Now the remainder of the resist can be removed chemically.

The wafer is then exposed to an atmosphere of a gas that will deposit boron atoms on the surface. Heating accelerates diffusion of the boron into the wafer. Most dopants diffuse very slowly in oxide, so the oxide layer of about 1 μm thickness protects the rest of the silicon. Ga and Al diffuse well in oxide, so they cannot be used in this process. P, As or Sb can be used as dopants if an N region is desired. Finally, the oxide layer can be removed, and the process repeated to form areas with other characteristics. This use of the oxide as a diffusion mask is one of the great advantages of silicon. The wafer material we start with is called the *substrate*.

Interconnections can be made with Al metal. The wafer

is given a thick coat of oxide, 1 or 1.5 μm thick, and then a layer of Al about 0.5 μm thick is evaporated all over. A mask is prepared, and resist exposed through it, just as for diffusion. The unexposed resist, corresponding to the areas where we do not want Al, is washed off. The unwanted Al is then dissolved in dilute acid, and all the electrical connections are made. The resistance of a conductor of length l and width w is $R = 0.05(l/w)\ \Omega$, where the resistance "per square" of the Al is 0.05Ω (this is the resitivity times the thickness). Another conductor is heavily doped polycrystalling silicon, called *poly*.

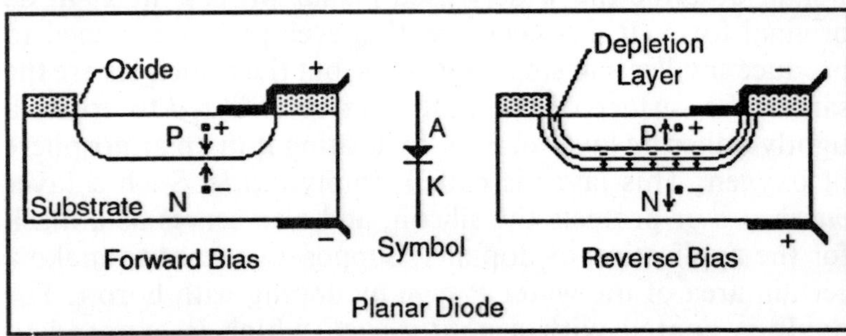

Planar Diode

As an example of what we can now do, consider a P-type region diffused into an N-type substrate. A side view is shown in the diagram. At a certain depth, there is a sudden change from P to N, called a *PN junction*, without any disruption to the crystal structure. If we make the P end positive, and the N end negative, an electric field points from P to N. It drives the holes toward the junction in the P region, and the electrons toward the junction in the N region, so the current flows in the same direction, P to N. When the holes meet the electrons at the junction of the P and N materials, they recombine, while new holes are created at the P terminal and new electrons at the N terminal (it is possible to make connections so that this takes place). The PN junction is *forward-biased* and carries current easily. The arrow in the diode symbol points from P-type to N-type, in the direction of easy current flow. The P-type material is the anode, the N-type the cathode.

Now reverse the polarity, making the P region negative and the N substrate positive, as shown on the right-hand side of the diagram. The electric field that results drives the holes and electrons away from the junction on each side. This creates a *depletion layer* with no charge carriers of either sign. Current cannot pass through this layer. As the carriers move away from the junction, they uncover the negative acceptors and the positive donors, so an electric field is created that cancels the applied field. The donors and acceptors are, of course, fixed in the lattice and cannot move. The current cannot continue, for there is no way to get new holes at the junction for the P material, and no way to get new electrons for the N material. The semiconductor now acts like an open circuit. The voltage across it is equal to the applied voltage. That is, the PN junction *rectifies*.

Diodes can be (and usually are) made by the planar method. We can start with an N-type wafer, then diffuse a P-type region into it. The substrate becomes the cathode, while the P region becomes the anode. A thousand of these can be made at one time on a wafer, and cut apart afterwards. Mass production is a feature of the planar process. Individual diodes, or whole microprocessors, can be made in large numbers with processes common to all, greatly reducing the expense.

When we apply forward bias and current flows, holes and electrons must recombine in large numbers near the PN junction. Holes are driven into the N region, where they become minority carriers, and electrons into the P region, where they likewise are in the minority. Recombination takes place to restore equilibrium. In silicon, the energy released by recombination is given up to the lattice. In some other materials, such as gallium phosphide, the recombination also can take place with the emission of light, at a wavelength corresponding roughly to the band gap, according to Bohr's formula $h\nu = eV$. Now we have a light-emitting diode, an LED. The wavelength emitted is given roughly by $\lambda = hc/eV = 1240/V$, if λ is in nm and V is in volt. Gallium arsenide, GaAs, has a band gap of 1.35 eV, so it will emit around 920 nm, in the infrared. Gallium phosphide, GaP, with a band gap of 2.24 eV, will give about 553 nm, a green light at the maximum spectral sensitivity of the eye.

If there is a reverse bias instead, the current through the

PN junction is very small. The area from which the charge carriers have been swept, and in which there is a strong electric field, is called the *depletion layer*. If light passes through this region, there is a good chance that it will create an electron-hole pair, and the two charges will be quickly swept away before they can recombine again. A *photocurrent* results, and the PN junction is now detecting light; it is called a *photodiode*. The maximum wavelength that will make charge carriers is given by the same formula as for emitters. Silicon, with a band gap of 1.12 eV, will be sensitive to wavelengths shorter than 1110 nm. A GaAs emitter, at 920 nm, will be a good match for a Si detector, at 1110 nm, and this pair is used often in practice, as with TV remote controls.

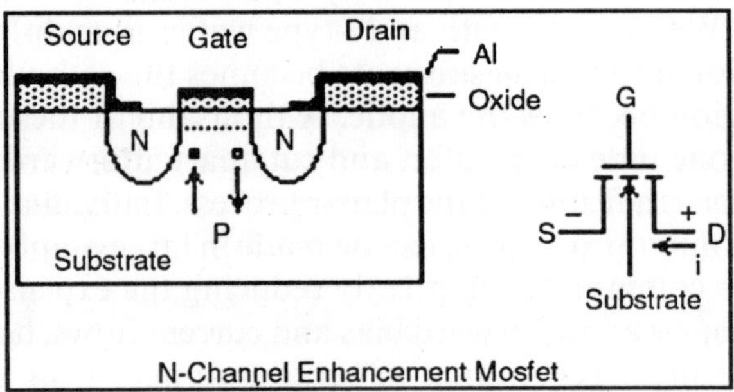

N-Channel Enhancement Mosfet

A simple device easily created with planar technology is the field-effect transistor, or FET. This device controls a current by means of a voltage applied to it. Since no current is drawn by the control circuit, no power is required to operate it, and the power gain is effectively infinite. A side view is shown in the diagram, together with the circuit symbol. We start, for example, with a P-type substrate with a resistivity of about 5 Ω-cm, and diffuse two heavily-doped N regions into it. These regions are the main terminals of the device, called *source* and *drain* though they are actually alike and either can be chosen as the source. Whichever way we apply a voltage between them, no current will flow because there will be a reverse-biased junction at one electrode or the other. Conventionally, the drain has the reverse-biased junction.

A thin layer of oxide, perhaps only 0.1 μm thick, is deposited between the two electrodes, and on top of it is evaporated an Al *gate* electrode. If we apply a positive voltage with respect to the source to the gate, an electric field appears under it in the P-type substrate. This field drives away holes and attracts electrons to the thin layer just under the gate. At some voltage, electrons will predominate and the type of the region will be *inverted* from P to N. This means that the reverse-biased junction at the drain will disappear, and this will allow current to flow from drain to source (electrons from source to drain, as the names indicate). The transistor has now "turned on" and conducts. This control is exerted purely by the voltage at the gate electrode. The device is usually called a MOSFET, whether metal or polysilicon is used for the gate electrode. Microprocessors and memory chips are all MOSFET's.

Field-effect transistors were sought by workers even before there was silicon or any other solid-state device, but surface effects and impurities always frustrated their efforts, and the most they could devise were rectifiers. The galena and pyrites crystals of crystal sets are solid-state rectifiers. When how to use germanium (and later, silicon) for diodes became clear in the mid-20th century, transistor action on a different principle was discovered in 1948. The first devices used point contacts, like the crystal sets, but soon the much superior junction devices were invented. With a crystal set, you poke around with the catwhisker trying to find a crystallite that is doped just the right way to be P-type.

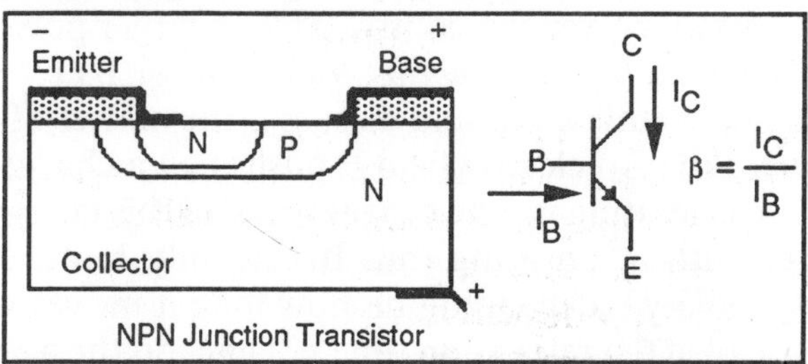

NPN Junction Transistor

A junction transistor consists of two N-type regions separated by a very narrow P-type layer, called the *base*. A planar junction transistor is shown in the diagram. In a point-contact transistor, the N regions are metal whiskers touching the base close together, and act in the same way that we shall describe. By making the base positive with respect to one of the N-type regions, which we shall call the *emitter*, we can cause a large current of electrons to move from the N-type emitter into the P-type base.

Meanwhile, we have made the other N-type region, the *collector*, positive with respect to the base, so that no current would normally flow, because the junction is reverse-biased. Emitter and collector correspond to the source and drain of an FET. The essential thing is to have a forward-biased emitter junction very close to a reverse-biased collector junction.

In the collector junction there is a strong field pushing electrons into the N-type collector region. There are usually no electrons to speak of in the P-type base, so there is usually no collector current, just a small leakage current from thermally generated electrons.

However, when we forward-bias the emitter-base junction, all of a sudden there are lots of electrons milling about the P-type region, and they are much more likely to fall into the collector than anything else. Some do recombine, and this requires a small current into the base to supply the holes lost this way. Most, however, just continue on into the collector. Therefore, the collector current is proportional to the small current that flows into the base, but much larger, by a factor of 100 or so, called the transistor's *beta*. This *transistor action* provides a very large power amplification, in which a small power can control a much larger power. This device is called a *bipolar junction transistor*, or BJT.

One final detail we should discuss is the matter of charge carrier *mobility*. An electric field E pushes on a charge e with force eE, accelerating it. The charge loses all this energy when it collides with a vibrating ion. It can only build up to a terminal velocity v depending on how long it moves without colliding, called the relaxation time τ. Then, on the average, $v = (e/m)E\tau$, proportional to E, by Newton's second law. The mobility is the constant of proportionality μ, so that $v = \mu E$, and its units are cm/sec/V/cm. Once we have the mobility, we can calculate the conductivity of a semiconductor, since the current density $j = nev = ne\mu V$, where n is the charge carrier density. The conductivity is then $\sigma = ne\mu$, which is a measure of the carrier density. In silicon, electrons have mobility 1900 cm^2/V-s, holes 500 cm^2/V-s.

The P-type substrate for an FET mentioned above was said to have a resistivity of 5 Ω-cm. Using $\sigma = ne\mu$, we find that this corresponds to a doping density of 6.25×10^{16} cm^{-3}, which is rather lightly doped. There are 5×10^{22} silicon atoms per cc, so only one in a million is replaced by a boron. A hundred times this is a rather high doping level.

SILICON AND MINERALS

Oxygen and silicon alone constitute 74% of the mass of the earth's crust. If the composition of the whole earth were known, they would probably be even more predominant. Approximate percentages are O 46%, Si 28%, Al 8%, Fe 5%, Ca 4%, Na 3%, K 3%, and Mg 2%, for a total of 99%. All other elements make up the remaining 1%. It is hot within the earth, but not molten in the usual sense. The contents of the mantle grind around in slow convection, while the foam is scraped off the top to make the crust. The crust is only 33 km thick on the average, and is the only part we know directly. The lighter metals, Al, Na and K, are concentrated at the top, in rocks called the "sial," while the heavier metals, Fe, Mg, Ca are concentrated at the bottom, in the "fema." Silicon predominates at every level. The lighter rocks have been crushed down into mountain roots by continental drift, where they may form molten blobs called *magma* where a lot of rock chemistry occurs. Erosion brings these areas to the surface as *batholiths* of granitic rocks, surrounded by the tortured writhing of gneisses, showing us what once happened deep in the crust.

All these rocks are mainly silicates, which form perhaps 80% of all minerals. The simple minerals, many of which are metallic ores, like galena and pyrites, are rare and have been concentrated in local deposits. By far the majority of minerals have crystallized from liquid and solid solutions of mixed silicates, so what we observe is the result of a process that produces infinite variety, controlled by a few basic principles. I will try to explain the principles, with a few examples of how they give rise to the observed rocks, without getting lost in the confusing variations that make igneous mineralogy so complex and puzzling.

An *igneous* rock is one that has crystallized from a liquid *magma*, or has been formed by diffusion and recrystallization below the melting point. The mantle of the earth (the thick layer between the crust and the core that makes up most of the volume of the earth) is not composed of melted rock, as was once assumed, but rather magmas are created locally by the grindings of plate tectonics, which also provides active fluids that dissolve and modify rocks. Old magmas are exposed by erosion as batholiths here and there, and their rocks have *intruded* surrounding rocks, dissolving and wedging into every crack. Rocks that show evidence of later modification by heat, pressure and the action of these active fluids are separately classified as *metamorphic*. Igneous and metamorphic rocks are quite distinct in structure from those that are composed of material weathered from pre-existing rocks, and deposited in the ocean or on dry land, where they have become cemented, hardened and consolidated into *sedimentary* rocks like sandstone, shale and limestone. Sandstone and shale are mostly silica, but the limestone represents the carbon dioxide that was once present in the atmosphere.

The basic structural element of silicates is the *silica tetrahedron*, formed by a central Si atom surrounded by four O atoms, each with an unpaired electron. This unit can be written SiO_4^{----}. It can form four bonds, covalent or electrovalent, with other atoms. This *orthosilicate* ion is stable and can actually exist, as in orthosilicic acid, H_4SiO_4. However, it normally forms covalent bonds with an adjacent Si atom, so that the Si atoms are connected by an oxygen bridge: -Si-O-Si-. This is a strong bond. This can be considered as formed from condensation of two -OH groups attached to Si atoms: -Si-OH + HO-Si- $\rightarrow$ -Si-O-Si- + H_2O. If every Si atom has four such bonds, then the structure can be represented by $(SiO_2)_n$. Normally, we write just SiO_2, but this molecule does not actually exist, at least in the solid, so the formula shows only the ratio of Si to O, which is 1 to 2.

If our melt contained only silicon and oxygen, then obviously silica would be the result on cooling. Comparing silicon and carbon, you might expect that the solid would have a diamond structure (as indeed crystalline elementary silicon does) but silica does not behave like this. There are more stable forms of more complex structures.

The familiar quartz crystals are of a structure called α-quartz, which forms hexagonal crystals. It is a very complex structure with over 100 atoms in the unit cell, in which the silica tetrahedrons form a kind of spiral. It is so difficult for the atoms to arrange themselves in this structure that molten quartz solidifies as a glass, called *fused quartz*. Using special crystal-growing techniques, α-quartz crystals can now be grown in the laboratory, and have replaced the natural crystals, which were growing rare in the required sizes. Nevertheless, it is not an easy thing. Natural crystals have required 100,000 years or so to grow.

Quartz is also found in crystal forms that are stable at higher temperatures. *Cristobalite* is cubic, and *tridymite* is hexagonal. Both have structures similar to those of ice, a curious thing because quartz was once considered petrified ice! Keatite, coesite and shishkovite are more recently discovered crystal forms, that do not occur in the normal course of events. The last two are found in meteor impact craters, where they were produced by the high pressures and temperatures. The crystal structure of quartz is a very complex matter.

Quartz is *birefringent* and *optically active*, properties used in the handling of polarized light. It is *piezoelectric*, meaning that distortion of the crystal produces electrical polarization, used in the *crystal resonators* that can control the frequency of electrical oscillations. The "quartz" legend on the face of a wristwatch means that there is such a crystal inside. It is also *pyroelectric*, meaning that heating produces electrical polarization, for which a use is yet to be discovered.

All these topics are worthy of exploration, but would require an article of their own devoted to the optical and electrical properties of quartz. Crystalline quartz is often colored by impurities, making a variety of attractive gems, such as purple amethyst (Fe^{+++}), yellow citrine (iron hydrates), pink rose quartz (Mn^{++}), and blue quartz (needlelike colloidal particles of TiO_2). Microcrystalline quartz also occurs in many attractive forms, such as variegated agate, red jasper, chalcedony, layered black-and-white onyx, red-and-white sardonyx, green chrysoprase, green with red drops heliotrope or "bloodstone," and orange-red carnelian, a variety of chalcedony including limonite and hematite, which are iron oxides.

Opal is a curious form of silica that includes some water. Its formula is usually written $SiO_2 \cdot nH_2O$. It consists of small, regular spheres of cristobalite, cubic silica, about 300 nm in diameter, cemented by hydrated silica gel. The fact that the size of the spheres is comparable to the wavelength of light, and that they are arranged regularly, creates interference that causes an attractive play of colour. Mother-of-pearl, or nacre, secreted by oysters and abalone, has a similar iridescence and colour, but is calcium carbonate, layers of thin aragonite films. However, a grain of silica is often the irritant at the centre of a pearl. Precious opal is white or black and has a good play of iridescent colors, but common opal may be less showy and lack iridescence. The best comes from Australia. Fire opal from Queritaro is red, but not iridescent. Siliceous sinter deposited at hot springs, geyserite, petrified wood, and diatomite, which will be discussed in the next section, are composed of opaline silica. Opal may deteriorate on exposure to air due to loss of water.

The natural magma, however, also contains aluminium Al and other cations, such as postassium K. If an extra electron is given to the Al atom, it can form four tetrahedral bonds just as silicon can, and we can make alumina tetrahedra $(AlO_2)^-$ that are very similar to the silica tetrahedra, except that they have a (-) charge. These alumina tetrahedra can make covalent bonds with silica tetrahedra, or with other alumina tetrahedra. Anything we can build with silica tetrahedra can be built with a mixture of silica and alumina tetrahedra, provided only that we neutralize the negative charge by adding the same amount of positive charge. This can be done quite simply by the unit ($KAlO_2$) which substitutes exactly for the unit (SiO_2). A phase with the constitution $(KAlO_2)(SiO_2)_3$ commonly crystallizes out. The formula may be written in the less evident form $KAlSi_3O_8$. This mineral is *orthoclase*, a potassium feldspar commonly found in granites.

Granite is formed from a melt (just our name for the original substance; it does not have to have been actually molten) rich in quartz, which at high temperatures is a homogeneous solid solution. When it cools, the components are no longer miscible, and it separates out into distinct phases. These may be quartz, and the potassium feldspar we have just mentioned.

A granite is mainly feldspar, perhaps pink in colour, with large masses of quartz embedded in it. Granites are a *plutonic* or *abyssal* rock, meaning that they have crystallized in a leisurely manner deep in the earth, so that the individual crystals can grow large enough to be easily recognized by the eye. The granite is then *phanerocrystalline*. If it crystallized in intrusive sills and dikes, that is in *hypabyssal* conditions, the crystals are of microscopic size, and we have a *rhyolite*. If it had partially crystallized before being intruded, it would be a *granite porphyry*, of large crystals in a microcrystalline groundmass, or a *granophyre* if the groundmass also had visible crystals. If it flowed out over the ground, and cooled quickly, under *extrusive* conditions, it would be a black glass, *obsidian*. If it were blown into the air as a foam, it would form *pumice*. Geologists are very interested in the appearance of the rock and its crystal size, greatly multiplying the descriptive names that are used for the same mixture of minerals. Mineralogists like to have one rock, one name for ease of labelling their collections, and look askance at the collective terms that are so useful to us.

If the magma also contained Ca and Na, the units ($CaAl_2O_4$) and ($NaAlO_2$) might also appear in the solid phases. The radius of the Ca^{++} ion is 0.099 nm, and the radius of the Na^+ ion is 0.095 nm, while the K+ radius is 0.133 nm, considerably larger than either. It is clear that the calcium and sodium units could substitute for one another without anyone's being the wiser, but that potassium would have a squeeze in a calcium-sodium structure.

This, in fact, does happen, and the potassium feldspars, and those with sodium and calcium, form separate phases. $(NaAlO_2)(SiO_2)_3$ is called *albite*, while $(CaAl_2O_4)(SiO_2)_3$ is called *anorthite*.

There is everything in between, as well, depending on how much sodium and how much calcium happens to be around. This series is called *plagioclase* feldspar, and names have been given to the compositions commonly seen. The granite, then, probably contains some plagioclase feldspar as well as the orthoclase.

The magma probably also has some iron and magnesium in it as well. These ions crystallize in a phase that is usually *biotite mica*, small black, brown or dark green scaly plates. "Mica" is Latin for crumb, referring to these flaky bits.

The minerals we have so far described have been *tectosilicates*, 3-dimensional structures like geodesic domes, with the K, Na and Ca of the feldspars inside. The silicon:oxygen ratio in tectosilicates is 1:2, as in silica, and all oxygens are shared.

The net formal charge is 0, so aluminium tetrahedra must be added to include cations. Magnesium and iron ions would rattle around in these spaces, so a new tighter structure is favored. Imagine the silica tetrahedra as sitting on their bases in a plane, and rearrange them in your minds as forming hexagonal rings, the vertices having condensed together.

The whole plane can be tessellated in this way, and the composition is $(Si_2O_5)^{=}$.

If magnesium is used to get neutrality, then the unit is $(MgSi_2O_5)$, and the layers are electrically neutral. This is very similar to the soft mineral talc, which adds a $Mg(OH)_2$ for each two of the above units. The mineral comes apart in thin sheets, which is typical of talc. Silicates based on sheets of tetrahedra are called *phyllosilicates*, from the Greek "phyllon," "leaf."

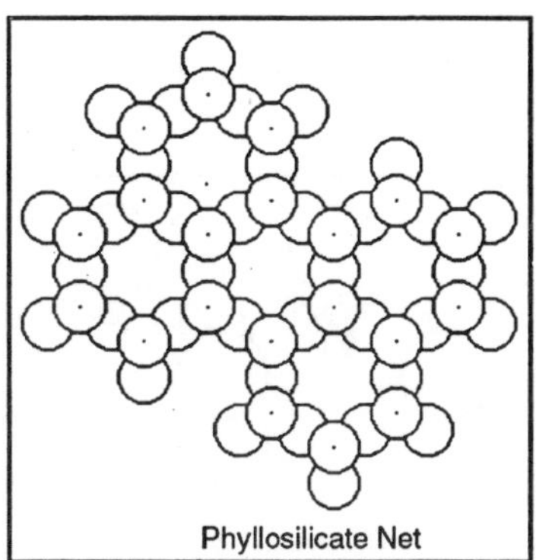

Phyllosilicate Net

The phyllosilicate net is shown at the right, and should be considered as extending in all directions in the plane. All the silica tetrahedra are shown with one orientation, but the unbonded oxygens could point to either side, and could be random or ordered.

The black dots show the locations of the silicon atoms. In a ring, there are 6 silicons, 12 oxygens belonging exclusively to the ring, and 6 shared oxygens. The silicon:oxygen ratio is then 2:5 for this structure. It is also possible for some of the silica tetrahedra to be replaced by aluminium tetrahedra. For all silica tetrahedra, the net charge of a ring is -2, which must be balanced by cations in the crystal structure. Each aluminium tetrahedron adds an additional -1 to the formal charge. There is electron resonance to give additional stabilization, so the negative charges are not associated with any particular unbonded oxygens.

Our magma, however, must take care of its aluminium and iron somehow, but is low in magnesium, so the micas form. The micas include alumina tetrahedra in the construction of the sheets, which have a negative overall charge. They are held together by positive ions between the negative sheets. The micas still come apart in flakes, splitting apart along the weaker interfaces between sheets, but the sheets are harder and more coherent than those of talc. Now we finally have a complete granite: mostly pink orthoclase, with some plagioclase, free silica in transparent crystals, and dark flecks of biotite or clear scales of muscovite. Rocks of this general composition are called *felsic*, from feldspar and silica, their principal components. Felsic rocks contain more than 65% silica.

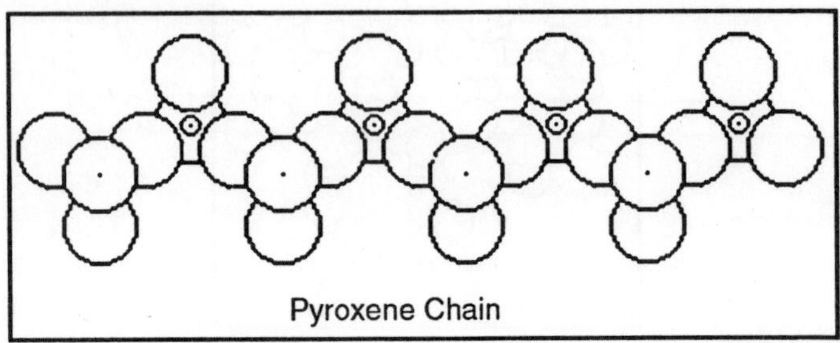

Pyroxene Chain

Silica tetrahedra can also be put together in chains. Such silicates are called *inosilicates*. A single chain, with silica tetrahedra connected at one point, has the composition $(SiO_3)^{-}$, with a silicon:oxygen ratio of 1:3. Such a chain is shown at the left. There are plenty of opportunities for cross-linking to make a solid crystal, but the chains still tend to cleave easily along their length.

These minerals are called *pyroxenes*, of which a simple example is *enstatite*, $(MgSiO_3)$. This formula looks like just magnesium silicate, a simple salt, but it is not—it is a crystal with this unit. Enstatite is a common mineral, cleaving easily in the direction of the chains, and sometimes found in fibrous masses. It is hard, heavy and green or olive-colored.

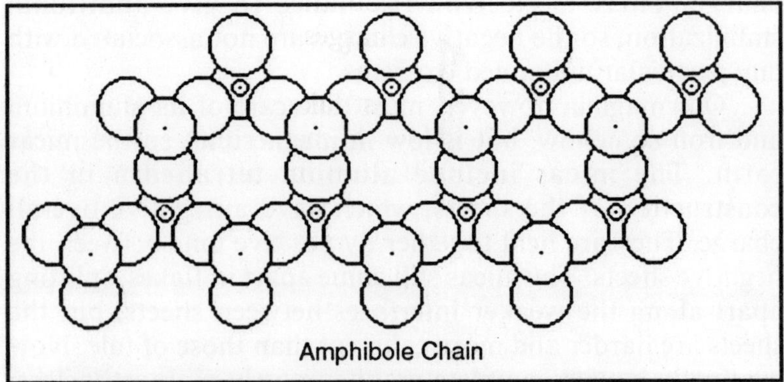

Amphibole Chain

Corresponding side links of the chain can be condensed, forming one double chain, as shown in the figure at the right.

Minerals with this structure are *amphiboles*, also inosilicates, and the unit is $(Si_4O_{11})^{-6}$, with silicon:oxygen ratio 4:11. In *tremolite*, the negative charge -12 of two chain units is neutralized by $CaMg_5$. If the Ca^{++} is replaced by Fe^{++}, the result is *actinolite*.

In either case there is an extra $Ca(OH)_2$ in the molecule. Tremolite and actinolite are members of a series, like the plagioclase feldspars. Actinolite is green and fibrous, giving us *nephrite jade*. *Jadeite*, the other form of jade, is a pyroxene in which NaAl neutralizes two chain units.

Tremolite is often fibrous, and was the material originally called *asbestos*. In Greek "asbestos" means "unquenchable, irrepressible" as an adjective, and "unslaked lime" as a noun, which seem exactly opposite to the characteristics of the mineral fibre.

It was a wonderful substance, that could be picked apart into a fibre of which a cloth could be woven that could not be ignited or damaged by acids. It was later found that the metamorphic rock serpentine (see below) also has a similar fibrous phase called *chrysotile* that made a better asbestos. Unfortunately, chrysotile dust causes a virulent form of cancer, a fact discovered late in the 20th century, after asbestos, a very useful substance, had been widely used. Tremolite is not dangerous.

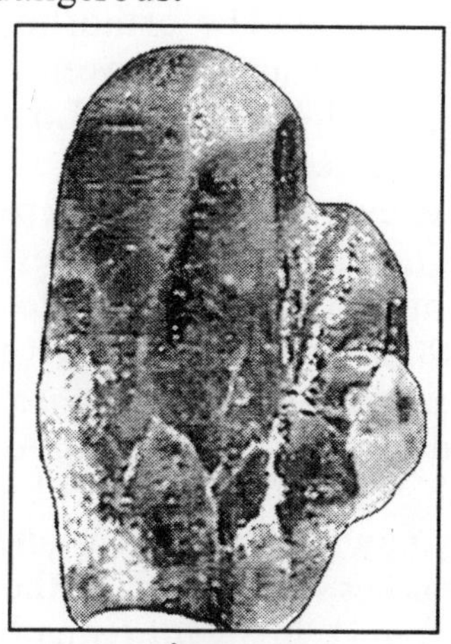

Orthosilicates are an important type of silicates. The

formulas look like salts of orthosilicic acid, Mg_2SiO_4, but the crystals are not, of course. They consist of silica tetrahedra, $(SiO_4)^{-}$, in which the tetrahedra are joined not by condensation at the corners, but by cations that hold the oxygens of two silicons in each hand.

If the unit is $(MgSiO_4)$, the crystal is *olivine*, which may be the most common mineral of all on earth, if the earth's mantle is indeed made of it. Fe^{++} can easily substitute for Mg^{++}, since their radii are 0.083-0.075 nm and 0.078-0.065 nm, respectively. Again we have a series like that of plagioclase. The magnesium end is *forsterite*, the iron end *fayalite*. Olivine is the generic name for the whole series of minerals, and its crystal lattice is orthorhombic. A specimen of transparent forsterite with the characteristic colour is shown at the left (image ©Amethyst Galleries). The green is due to ferrous iron.

The forsterite end gives us the beautiful gem *peridot*, of deep olive-green colour. Stones with a yellow-green colour are called *chrysotile*, not to be confused with its anagram *chrysolite*, which is asbestos. The hardness is nearly Mohs 7, equal to quartz. Optically, it is a positive biaxial crystal, with indices of refraction 1.6359, 1.6507, 1.6688, greater than that of beryl (1.58) or quartz (1.5442ω, 1.5533Å), about that of ruby and sapphire (1.77), but much less than that of zircon (1.92) or diamond (2.42). The orthorhombic crystals are distinctly pleochroic. The angle between optic axes is 2V = 85° 16'. The Geological Museum in London once had on display a perfect, faceted peridot of 136 carats, the most beautiful gem I have ever seen. Most gem peridots are at most 30 or 40 carats. The pleochroism was evident, and gave a subtle variation of colour. Gem peridot has been found predominantly in St. John's Island in the Red Sea, in a deposit of *dunite*, a mysterious rock from the deeps that is mainly olivine. It is named after the Dun Hills in Australia, I am told. I have not been able to find the Dun Hills in my atlases. St. John's Island may now be Shadwβn Island at the north end of the Red Sea. Another source says it is the island of Zebirget off the port of Berenice in Egypt (equally ex-atlas). Other sources are said to be Burma, Queensland, Brazil and the Navajo country of New Mexico and Arizona. Much modern peridot is said to be reworkings of earlier stones. Artificial forsterite has been made as a high-temperature refractory material. It is made by fusing olivine rock with magnesia, to bring the magnesium content up to that of forsterite.

Olivine has been detected in interstellar dust by means of an infrared absorption near 9.7 μm. This was done by Krδtschmer in 1979. This helps the case for olivine as a primary constituent of planets.

Classification of Rocks by Silica Content

Name	Silica
Felsic	>65%
Mesosilicic	52-65%
Mafic	45-52%
Ultramafic	<45%

The minerals we have just discussed crystallize from a melt that is poor in aluminium and alkali metals (otherwise feldspars would form instead of pyroxenes), but richer in magnesium and iron. Instead of granite, we get gabbro, containing plagioclase and pyroxene. Its hypabyssal equivalent is basalt, instead of rhyolite. The feldspars give light-colored rocks, the pyroxenes dark-colored rocks. The classification of rocks by silica content is explained in the table at the right. The percentages are calculated for the oxides; for example, silica is calculated as SiO_2. Forsterite is considered as $2MgO \cdot SiO_2$, so its percentage of silica is 43%. Olivine is an ultramafic rock. Albite, a plagioclase feldspar, is 69% silica, and so is felsic. The classification mainly refers to rocks that are an assembly of minerals, rather than to the minerals themselves.

Olivine and dunite are found in characteristic assemblies of rocks called *ophiolite suites*. In Greek, "ophis" means snake, and this is an allusion to the typical occurrence of *serpentine*, a metamorphic rock that is produced from olivine, pyroxene and similar mafic and ultramafic rocks by heat and pressure, in the assembly. Serpentine occurs as chrysotile, antigorite and lizardite, and is commonly greenish like olivine. We have already talked about its fibrous form, chrysotile. These rocks are found at spreading centers on the edges of continental plates, where basic magma from the mantle is brought to the surface. The mid-Atlantic ridge is one such place, Cyprus is another, and the Red Sea overlies another line of spreading that has separated Africa and Arabia, and continues down the Rift Valley of Africa. The occurrence of lizardite on the Lizard Peninsula in Cornwall, and dunite in Australia, may be relics of earlier spreading centers. Before plate tectonics, the origin of these rocks was a subject of great perplexity.

Silicate minerals show clearly that what is important about an atom are only its size and charge, and that they are assembled into compounds any way that gives the least free energy. What results depends on the constitution of the melt. There is no calcium analog to forsterite. Calcium, at 0.099 nm, seems to be too much larger than magnesium at 0.065 nm. Zinc, on the other hand, has a radius of 0.074 nm, which is close enough, and there is an analog to forsterite, which is the fluorescent mineral *willemite*. We have series of minerals in which one ion replaces another, Ca-Na and Mg-Fe for example, making minerals that are carefully distinguished by the mineralogist, but are really the same thing to a good approximation. Zn fits in with Mg and Fe, but it is too rare to make much of an impact. Oxygen and fluorine negative ions are the same size, but differ in charge. F often substitutes for OH, notably in micas. SiF_4 can't make macromolecules, so Si always prefers the oxygen.

SILICA AND THE SEA

Life burgeoned in the sea, but this world is still largely unknown to the general public, except for the large sea life they plunder and eat. The occupants of the sea are the *plankton* that float passively at the surface, the *nekton* that swim in its volume, and the *benthos* that lie on the bottom. Plants exist only among the plankton, since they are dependent on sunlight to drive photosynthesis, constituting the *phytoplankton*, the "plant-plankton." They are the source for all the food in the ocean, which covers 70% of the area of the planet. All other life eats them, smaller things that eat them, and whatever drops to the bottom. Phytoplankton is largely ignored in ecological arguments, though it must play a major role not only in biology but also in atmospheric and oceanic science. It is well out of sight, and hence out of mind.

Plants contain chlorophyll, contained in organelles inherited from earlier, simpler organisms who could not survive in the new oxygen-rich atmosphere created by the forms of life that released it to the atmosphere as a waste product, and poisoned themselves. The chlorophyll manufactured complex compounds of carbon from which energy could be derived by oxidation. Animals ate what they could find of these substances. Movement and sense organs were stimulated by the search for food. Much of this life was single-celled or *acellular*, microscopic and soft, not preserved in the fossil record.

Cellular plants usually made use of strong, rigid cell walls constructed from cellulose to give them the ability to grow large and complex. Animals used chitinous or cartilaginous substances for the same purpose, as framework to grow on, or as protective enclosures. Then animals found they could survive more comfortably by eating smaller animals that grazed on phytoplankton, and spare themselves the trouble of filtering huge amounts of water. The response to predation was the favoring of armored coverings, and none proved better than calcium carbonate, which could be extracted from the sea water (the richness of the atmosphere in carbon dioxide led to thick deposits of calcareous sediments, and sea water was full of carbonates).

This occurred at the beginning of the Cambrian era, and suddenly the fossil record is full of calcareous shells, which were easily preserved in conditions favorable to calcareous sediments. It should be remembered that the presence of a fossil is evidence of the existence of an organism, but the absence of a fossil is by no means evidence of the absence of an organism.

The *foraminifera*, of which *globigerina* is an example, were acellular animals that chose a carbonate skeleton to grow in and around. Their fossils are found beginning in the Ordovician. However, another protozoan that chose a siliceous skeleton, the *radiolarians*, is found from the Cambrian on.

These fossils, or *tests*, look like spiky balls full of holes, from which the pseudopodia of the animal projected. The *porifera*, sponges, were multicelled animals with undifferentiated cells and no sensory organs. Some chose to grow on siliceous *spicules*, which also left a fossil record beginning in the Cambrian.

It is easy to call these animals "primitive" as if they were just an inferior stage of development, but it should be remembered that they still exist today, which shows that they are well-adapted to their environment, and very successful.

Now we come to our main subject, the *diatoms*. Diatoms are microscopic plants, algae, that belong to the *phytoplankton*, plants that float wherever the sea takes them. In the Jurassic, they appear for the first time in the fossil record, since they then developed tests, or shells, or valves, or more precisely *frustules* (Latin, "little morsels"), of opaline silica.

The "opaline" means that it contains some water in its structure. They probably had always been around, since they are essential to life in the sea. There are other algae that help diatoms with photosynthesis, but diatoms are the primary sea plant, living in astronomical numbers in oceans and fresh water.

Every drop of sea water (and many drops of fresh water) will contain diatoms. They prefer higher latitudes, upwelling currents and continental shelves, and float at depths less than 70 metres, where they have sunlight and nutrients. The land of the earth produces about 40 million tons of carbohydrates from photosynthesis, but the sea produces 80 to 120 million tons, mainly due to diatoms. Beneath a square metre of surface of the Gulf of Maine, for example, there are from 7 to 8 *billion* diatoms. One diatom can have 100 million descendants in a month. They are astonishingly prolific and efficient, making the sea as productive as an equal area of tropical rain forest. They are one of the dominant factors in the CO_2 economy of the world, though seldom regarded.

Diatoms are the foundation of the food chain in the sea. They are eaten by copepods and other small zooplankton. The small crustacean *Calanus* grazes on diatoms, herring eats the crustaceans, and people eat the herring. Problems with the diatoms, calanus, or herring can interrupt the supply of fish. Herring was made extinct in the Baltic by overfishing long ago. This probably encouraged the crustaceans, who ate all the diatoms, and the whole chain collapsed. There is still no herring in the Baltic. Any economy of the sea must take the diatoms into account.

Although our main interest here is in silica-using diatoms, as far as photosynthesis in the sea is concerned we should not neglect the minute cyanobacterium *Prochlorococcus*, discovered as late as 1988, which probably accounts for half of the marine photosynthesis. I have no idea how these tiny cells (500-700 nm diameter) are eaten, since they would be difficult even to filter unless clumped, but they are no doubt the foundation of an important food chain. These cyanobacteria, one of the earliest forms of life, are also the most successful, it would appear.

The frustules of a diatom are in two parts, like a pillbox and its lid, that fit into one another, as shown in the diagram below. They are not hinged like the shell of a clam. This allows the plants to multiply by cellular division. Half of the protoplasm sticks to one half of the frustule, half to the other, and they come apart. Each half grows a new part of a frustule that inserts into the other, which forms a lid. The cells get a little smaller each time. Sexual reproduction creates spores that grow into a full-sized diatom to maintain the average size of the plant. This two-part frustule is the reason for the name "diatom" (Greek, "cut in two"). Although the protoplasm contains green chlorophyll, coloring agents in the shell give the plant a beautiful golden colour, so they are the family of the *chrysophyta*, "golden plants."

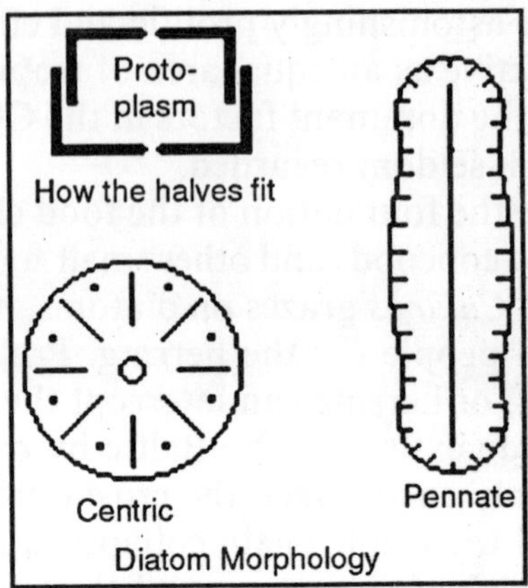

Diatom Morphology

The shapes of the frustules are either *penna te*, with bilateral symmetry, or most commonly *centric*, with radial symmetry, as illustrated in the figure. The pennate forms may even be *motile* to a degree, but all the centric forms are *sessile*. There is no reason for them to want to move, since sunlight and water are all they need.

Their decoration can be elaborate. The centric forms usually have a central pore and radial lines, while the pennate forms have a medial line and decorated edges. 5- and 10-fold symmetry is often seen.

These attractive forms can be examined under a microscope. It is estimated that there are 6000 to 10000 species of diatoms living at the present time, so they are eminently successful in what they do.

When the diatoms are finished with the frustules, they sink to the bottom, where they form *diatomaceous ooze*, if they are not dissolved. Diatoms like sunlight, salinity appropriate to their species, cool water, phosphate and nitrate nutrients, and, of course, silica. Sea water on the average contains from 0.02 to 4.0 ppm of silicon by weight, in water of standard 19% chlorinity. Diatoms are happier near the upper limit, where they can thrive. When frustules sink to the bottom, they take their silica with them, so it must constantly be supplied to the oceans. Where silica is plentiful, and the other requirements are met, diatoms will thrive, and make thick deposits of their shells. This rock is called *diatomite*, or "fossil flour," which has many uses, and is highly valued. Diatoms can also live in fresh-water lakes.

These lakes normally have very little silica, but vulcanism can supply volcanic ash and silica-laden hot waters, which run into the lakes and encourage diatoms. Something of this nature happened in the U. S. West, and especially California, in the Tertiary, which produced huge deposits of diatomite both in lakes, and in the ocean border. An arm of the sea reached in near Lompoc, just north of the Santa Inez Mountains, and with the nearby volcanic action, a thousand metres and more of diatomite was deposited in the Upper Miocene and Lower Pliocene, the largest deposit in the world.

Diatomite is 85-92% SiO_2, most of the rest being clay. When pure, it is brilliant white, but also can be yellowish or buff. Less pure rock is called *diatomaceous earth* or *Kieselguhr*. It has 75% porosity, more or less filled with water that can be driven out by heating. It can be quarried in blocks, that can be used directly for heat insulation. Its voids give it a high thermal resistance. Each cubic inch contains 40 million frustules or more, all of which are nicely sharp and all surface and pore space. The nature of the frustules is important in the application of the diatomite. It is a gentle abrasive, so it is used in silver polishes. In paint, it gives a matte finish as each frustule roughens the surface. It has excellent absorptivity for liquids. It was used as an absorbent for nitroglycerine, rendering it insensitive to shock, in Alfred Nobel's *dynamite*.

Its most important industrial use is probably as a filtering agent. A suspension in water is first passed through a screen, depositing a layer of diatomite on the screen. Then the suspension to be filtered moves through the prepared layer, leaving behind all but the finest matter. One source says that it even filters colloidal solutions, which pass through ordinary filter papers. This filtration is suitable for the "hard jobs" in clarifying liquids.

This is a suitable place to talk about the other siliceous rocks, most of which are *biogenetic*. The term *organogenetic* sometimes found is clumsy, looking more like "tool-made" than "made by life." The Cretaceous Chalk of England is a famous rock, mostly white calcium carbonate from foraminiferal tests. However, it also contains silica, now in the form of the familiar dark *flint* nodules, which typically occur in bands.

The thick beds were originally deposited as a siliceous chalk, the silica from opaline sponge spicules, and the chalk from foraminifera. Percolating waters replaced the opal in the spicules with calcium carbonate, and redeposited the silica as much less soluble silica concretions, called flints.

This separation into two purer constituents is not rare in sediments. They are roughly brick-sized elongated and rounded stones that remain covering the ground after the chalk weathers and erodes. They have been used like bricks, indeed, for walls and buildings, and even churches, since they are very durable.

Flint is deserving of special study, especially in the ways it has been used by man. Flint was prized by neolithic man for its conchoidal fracture, by which sharp, hard edges could be produced for tools and weapons. It was the earliest mined substance, since good flints for making tools are not common.

The flint mines in the chalk of Norfolk, in England, are famous. Flint-mining technology was later adapted for the mining of coal and metals. Flint was used for edged tools and weapons of all kinds, and great skill was involved in its shaping, called "flint-knapping." Egypt, a country poor in metals, used flint ploughs, showing how meaningless the term "neolithic" is as except as a rough cultural indicator. Flint was the origin of technology.

Flint and steel are well-known as fire-starters. When struck together, fat sparks are produced (probably from the steel, heated from impact and then burning in air) that are caught in tinder.

The glowing bit of tinder is put on dry grass and wood shavings, and coaxed into flame by gentle breaths. This was a much easier way to start a fire than any available alternative. The flintlock musket was invented around 1630, in which a spring-loaded flint hammer strikes steel to send sparks into the firing pan, where there is fine gunpowder, the *priming charge*, that ignites the granular gunpowder in the chamber. Flintlocks were still used in the early 19th century.

The United States Army cooled its heels in Piqua, Ohio waiting for flints in one phase of the War of 1812, on its way to Lake Erie. The "flints" in cigarette lighters are not flints, but an alloy of 70% *mischmetall* (50% cerium with 50% rare earth metals, mainly lanthanum), and 30% iron, patented by Auer von Welsbach in 1903, that gives off sparks copiously when struck by steel. Flint and steel are used in the "sparkers" used to ignite a gas torch with one hand.

Flint is a kind of *chert*, though chert is bedded, rather than occurring in nodules. Flints may be blackened on the outside, but when broken show a white centre that is almost pure silica.

Chert is also a biogenetic rock associated with limestones, but occasionally occurs alone in thick beds. Like flint, it is fine-grained or microcrystalline quartz or chalcedony. There has been a great geological controversy over the origins of flint and chert. Some hold that it is *syngenetic*, or created at the same time as the rock in which it is found, while others maintain that it is *epigenetic*, or formed later. It is probable that some flints began as silica gel blobs formed in the alkaline carbonate deposits, and so are syngenetic, but most flints appear to have formed later, as described above. Petrified wood and similar bodies where silica has replaced the original material, are clearly epigenetic. The mechanism of this faithful *metasomatic* replacement is unknown. As in so many other cases, both origins are probably involved under different circumstances. Bedded chert probably began as radiolarian or diatomaceous ooze, but time and metamorphism have erased the original fossils. It is difficult for the silica concentration to rise to the point where silica would precipitated out of solution. No such case is known, except for the siliceous sinter that forms at hot springs.

Radiolarite is like diatomite, except that the known examples are from the palaeozoic, and are hard and indurated, not soft and powdery like the Tertiary diatomites. Radiolarite has been used for whetstones and grinding stones.

SILICA AND CLAY

When a felsic rock weathers, water leaches out the potassium and sodium of the feldspars, which crumble into a soft phyllosilicate called *clay*, containing mainly aluminium, silicon and oxygen. The flakiness of clay minerals are the reason for the fissility of shales. Where for some reason they have not become aligned in layers, the result is nonfissile *mudstone*.

The crystals of silica are little affected by weathering, and remain as loose *sand*. Mixed with organic matter, *humus*, the result is *soil* in which plants can grow. Long weathering in humid tropical or semitropical areas may even leach out the silica, leaving only ferric and aluminium hydroxides, which are even more insoluble. This poor, reddish soil devoid of most nutrients is called *laterite* from its resemblance to brick clay (which, of course, it is not). Bauxite, the ore of aluminum, is a lateritic deposit consisting mainly of aluminium hydroxide. When basaltic rock weathers, the equivalent of clay is greenish *chlorite*, also a flaky mineral.

As our example of a clay, let's take the valuable *kaolin*, a pure form of clay that forms a slippery white mud, named after its original source in China. It is composed of six-membered rings of silica tetrahedrons, rendered electrically neutral by Al^{+++} and OH^- ions strategically arranged. Its formula may be written $Al_2Si_2O_5(OH)_4$, or $Al(OH)_3 \cdot HAlSi_2O_6$, or even $Al_4Si_4O_{10}(OH)_8$, which shows the Si to O ratio of 4 to 10 typical of the phyllosilicates. A "hydrated aluminium silicate" is a good description. It is, at any rate, a layered silicate like the micas. Clay usually is a complex mixture of related compounds with numerous impurities, but kaolin is a good representative of the family. Kaolin is called "pottery clay" because it is plastic, but it is mostly used for paper sizing, and only a minor fraction is used for pottery. Large amounts come from the weathered granite of Cornwall in south western England, where the clay separated in purification makes tall, conical white hills.

An exotic clay is *bentonite*, found in the Upper Cretaceous Benton Formation near Denver, and also in northeastern

Wyoming and other places. It is formed from the weathering of volcanic ash (whose major ingredient is feldspar), and occurs in irregular lenses in formations near the Cretaceous-Tertiary boundary in this region. Its major constituent is *montmorillonite*, a common clay mineral similar to kaolinite, in a colloidal form. Montmorillonite is a non-plastic clay with a maximum particle size of 1 μm.

Its distinction is that it has a great appetite for water, of which it is starved during its long dry burial in impermeable shales. When water can get to it, it will swell faster than a thirsty horse in a pond. It can adsorb 10 to 30 times its volume, and swell accordingly, the water coating the phyllosilicate sheets. A large amount thrown into water makes a gummy mass that is good for plugging leaks, since it is nearly impermeable.

It was used for this purpose during the construction of the Grand Coulee Dam in 1937, when a water leak threatened to undermine some important towers. Bentonite was forced into the sand under pressure, reducing the leak from 30,000 gpm to 250 gpm almost immediately. This makes it useful to plug leaks while drilling oil wells, but a rather unsatisfactory foundation for a house or runway. However, undaunted brave people keep building houses and airports on bentonite. Denver International Airport is built on bentonite in a tornado alley, but so far, so good.

Montmorillonite is also know as *fuller's earth*. It was used in the process known as *fulling*, which was the cleaning of wool from fats, performed in a slurry of montmorillionite by pounding. Now that good detergents are available, other methods are used. Cats like to bathe in montmorillonite if they can find a suitable clay to roll in. It cleans much more effectively than soap and water, and is much easier on the cat.

It was discovered that one could shape useful vessels from clay mud, and heat them in a furnace, or kiln, by the gases from a fire, and that they would harden into a porous stone and could be used for holding liquids and for cooking (if only by dropping hot rocks in them). The impurities in the clay form compounds, like slags, that melt at lower temperatures than the clay itself, and cement together the particles of the object. Kaolin is too pure to do this by itself. Pottery clay is carefully mixed from clay—the "right" clay—and sand, and perhaps other ingredients, to give the best product. Pottery that has been baked in this way is called "biscuit." It is usually red or buff colored from the iron impurities, which are oxidized in the heat of the kiln.

To give the objects a durable, nonporous surface, they must be *glazed* with some glassy substance melted so that it will flow over the surface and fill the porosity. This glaze must have a melting point lower than that of the slags binding the biscuit, so that the objects will not soften and sag when they are fired. Decoration may be applied to the biscuit object, and then the glaze, in the form of a liquid *slip*, is applied to the surface.

The object is fired again, to melt the glaze. The heat often changes the colors of the decoration as well as the glaze, as substances oxidize. Glazes are made from salt, $NaCl$, and metal oxides, such as those of tin or lead. Chemical changes occur when the object is fired. The salt becomes something else, so the glaze does not dissolve in water. The glaze can be clear, showing the decoration or the colour of the biscuit beneath, or may be colored. Borax with metal oxides produces good colored glazes.

A mix of white orthoclase, kaolin and quartz fuses at 1100-1300°C to a white, vitrified, translucent material called *porcelain*, long a Chinese trade secret. By using phosphates from burnt bones in a pottery clay, a similar material called *bone china* can be manufactured. The secret of making porcelain was finally decoded in Meissen, near Dresden, in the 18th century. Good porcelain is also made in Golden, Colorado using a local clay. Besides its use in fine tableware and ornaments, porcelain is useful in laboratory ware and as electrical insulators. Unlike glass, porcelain does not adsorb moisture that would spoil its insulating properties.

The clay may have a humbler destiny, as construction brick. Brick was originally made from clayey earth and a binder such as straw, mixed with just the right amount of water and pressed into molds. When the brick dries in the sun, the clay becomes hard and coherent. This makes a very satisfactory construction material, familiar as *adobe* in the Southwest. Such bricks were generally used in all parts of the ancient Mediterranean world until the first century BC. Egyptian palaces were made of them, as well as peasant's homes. In time, all melt into shapeless heaps of earth, since adobe is not preserved in the long haul. Homes in Greece and Rome were also of adobe, even substantial buildings. Only temples were in the expensive and hard-to-work, but very durable, stone.

Curiously, right now in Denver homes are built of plywood covered with tarpaper, which is inferior even to adobe, and are externally decorated with veneers that make them appear to be substantial buildings. They will rot and collapse even faster than the Egyptian palaces. They must be considered only temporary buildings.

A more durable material, rivalling stone in durability, can be made by baking the bricks in a kiln, like pottery. Baked brick was known in Babylonia, and used for limited purposes, but its widespread use in Europe was introduced by the Romans. The proper clay must be chosen, and it is mixed with water to make a stiff mud that is molded into the bricks. The Romans preferred a brick that was a relatively thin square, while we use a brick that is about 2"x4"x8". The bricks are fired, or *burnt*, in a kiln at a higher temperature than pottery, so they become strong, and usually red from oxidation of iron. Bricks for external use must be vitrified on the surface so they will not absorb water which will shatter the brick when it freezes.

Tiles are just bricks, but made in special shapes adapted for different duties. There are wall tiles for decoration, floor tiles for paving, and roof tiles for roofing. Ordinary bricks were also used for paving, even for streets, a service they performed very well and attractively. Corsicana, Texas, was called "The Bricks" on account of its paved streets. Bricks from Coffeyville, Kansas paved station platforms on the Santa Fe. Tall chimneys of brick kilns marked the outcrop of excellent brick clay in an arc from Oxfordshire to Huntingdonshire north of London. Brick is a warm, pleasant and human building material. Ignatz the mouse heaved brickbats at his admirer, Krazy Kat. A brickbat, incidentally, is half a brick, fitting neatly into the hand and of use in labour disputes and political demonstrations.

Bricks are laid using *mortar*, which is slaked lime, $Ca(OH)_2$, mixed with sand. Lime, CaO, is made by roasting, or "burning," limestone. A thin layer of mortar is spread to seat the bricks. The bricks absorb water from the mortar, and the rest of the water evaporates. The calcium hydroxide, some converted into calcium carbonate by the carbon dioxide in the air, crystallizes into a tangle of crystals that bind the bricks firmly by preventing relative sliding motion. Brick walls depend on weight for stability. They would not do on the space station. Mortar will not set in large amounts, like cement, only in thin sheets.

Denver was plagued by fires in its early, jerry-built history, but the abundance of good brick clay in the vicinity allowed Denver to rebuild and grow in brick. Many houses even had tile roofs. After the second world war, building codes were cut and shaped to maximize the profits of the builders, and this meant cheap wooden houses, as mentioned above. Unless the brick houses are torn down for space to build impressive million-dollar shacks, they will long outlast them. Much Roman brick survived its first use, and was reclaimed for use in later buildings. Some of Denver's recent wooden palaces were covered by an impermeable stucco-like sheathing. Water got behind it, and now the wood will rot out in less than a decade unless replaced.

The Romans also invented a substance like mortar, but which would set in large volumes and even under water, and made an inexpensive substitute for stone. This material, Portland cement, is made by roasting finely crushed limestone and clay in long, rotating cylindrical kilns. The limestone and clay are considered as bringing the constituents CaO, Al_2O_3, and SiO_2. These ingredients are mixed as a wet slurry in exactly the right amounts to produce the desired substances, which are: (1) $Ca_3Al_2O_6$, called Ca_3A; (2) Ca_3SiO_5, called C_3S; and (3) Ca_2SiO_4, called C_2S. The slurry is introduced at the top of the kiln, it soon is dried by the hot gases, and and tumbles downward until it fuses into *clinker* in the heat. The clinker from the kiln is ground to a very fine powder, since the reactions occur only on the surfaces of the particles, and everything inside is wasted.

It is called Portland cement because a Yorkshire bricklayer named Aspdin roasted limy road dust with river mud to make a powder that, when mixed with water and sand, would harden to a reasonable imitation of the excellent and expensive Portland stone from Southern England. He received a patent in 1824 for this discovery. Smeaton had used a less perfect cement at the Eddystone Rock lighthouse in 1756 by burning a clayey limestone. Cement does not set by drying. The water combines chemically with the cement.

When cement powder is mixed with water, reactions start immediately. Compound C_3A hydrolyzes to give calcium and aluminum hydroxides. Crystallization of these hydroxides gives the initial strength of the cement, up to about 24 hours from pouring. Meanwhile, the hydroxides are working on compound C_3S to form complex calcium aluminum silicates that increase the strength of the cement for up to about 7 days. C_2S works the slowest, giving the increase in strength from 7 days to 28 days, when the cement is considered to be fully set. Changing the relative amounts of the three ingredients controls the setting time and other characteristics of the cement. A little gypsum, $CaSO_4$ may be added to retard the initial set and give sufficient time to place the cement. Heat is given off when cement sets, and this must be allowed for so that the cement does not crack. A typical cement has an analysis of 60-70 CaO, 20-24 SiO_2, 3-8 Al_2O_3.

Cement is not usually used by itself, but in the cheaper but equally effective form of *concrete*, in which it is mixed with *aggregate*. Aggregate consists of coarse stones or gravel, and a finer material, sand. It is best for the sizes of the aggregate to be graded so that the voids are filled as well as possible. Concrete can be mixed on-site, but advantage is taken of the fact that if freshly-mixed cement is kept agitated, it will not set until the agitation stops. The crystals that would give the initial hardness are continually disturbed, and new material is continually supplied from the surfaces of the grains of cement. The distinctive trucks with the rotating tanks are evidence of this. If you stopped the rotation with the tanks full of concrete, you would have a tank full of solid concrete in a short time, but as long as it rotates, you can pour it.

An excellent construction material known as *cinder block* is made from cement mixed with rough cinders, and poured into forms. What is used now that coal fires are not as common, I do not know. Cinder was the slaggy material that resulted from reactions between sandy and limy material in the coal ash in a hot fire. The best was in small, sharp pieces that would interlock. It had many excellent characteristics, and was useful product that could be had for the effort of carrying it away. It made good footpaths, and much athletics took place on the "cinder track." Cinder block is much cheaper than brick, and lighter, and much more durable than wood. Why it is not more widely used is a mystery to me. It can be tarted up just as attractively as plywood and tarpaper can.

SILICA AND GLASS

I had thought of titling this section "Silica and Sand" but sand is usually silica, so it seemed redundant. Glasses are substances that are not crystalline, but tangled webs of long molecules. Silica loves to make long molecules, so it is used in glass for this purpose. A glass is a hard solid at lower temperatures, then softens at higher temperatures, usually above red heat, when it becomes soft and can easily be worked. At still higher temperatures, at white heat, it liquefies, and can be cast or a blob put on the end of a hollow tube so it can be blown into hollow shapes. Glass, like pottery, has been an outlet for artistic skills, and is of great beauty in addition to utility.

We think of glass as transparent, but this is a secondary characteristic. A glass must be quite pure to be transparent, or even translucent. One might wonder why glass can be so perfectly transparent, when it is a disorderly random tangle of silica chains. It is not hard to believe that an ordered crystal, like quartz or salt, could be transparent if the lattice spacings are smaller than a wavelength, and this can be shown mathematically. It is also true that a random structure can also be transparent, if the randomness is uniform on the average over distances of a wavelength or so. Gases, indeed, are an example of this, as is water. As Rayleigh showed, the blue of the sky is scattering from the *fluctuations* in density, not from the molecules as individuals. The disorder in glass is over distances less than a wavelength (about 500 nm), and so pure glass is transparent. Disorder over larger distances, or the inclusion of larger particles, makes a glass first translucent and milky, and finally opaque.

I have a beautiful glass bottle, an *unguentarium*, that is two thousand years old. It is 90 mm long by 15 mm in diameter, and weighs 6 g. It looks like a test tube with a somewhat bulbous end, and the glass is colorless. Clear glass was developed by the Romans, and was rather common. It was even used for windows, as we now use glass. Afterwards, houses just had holes in the wall in Western Europe, where the wind certainly did blow through, or at best oiled paper. Most Roman glass that is seen in museums is of common, crude objects that would have been cheap at the time and were discarded in trash, and so are easy to find. If you go to Colchester, for example, you will see in the Roman Museum there some fine Roman glass, that is clear, brilliant, symmetrical and very attractive, as good as that made today. Today, "pressed glass" made from cast glass is very beautiful, comparable to the much more expensive cut glass, but available for pennies.

One glass is made from quartz alone. Quartz (pure sand is used, not the large crystals) melts at 1800°C, and when it cools cannot figure out how to get back into the hexagonal quartz structure. This is a homogeneous and isotropic substance known as *fused quartz* or *quartz glass*. Fused quartz softens at 1665°C and anneals at 1140°C. These are very high temperatures, so fused quartz is difficult to work. Its density is 2.2 g/cc, its hardness Mohs 4.9 (not nearly as hard as quartz). Its bulk modulus is 5.3×10^6 psi, shear modulus $4/5 \times 10^6$ psi, Young's modulus 10.4×10^6 psi, and Poisson's ratio 0.16. In tension its strength is at most 7 ksi, but in compression it is very strong, 160 ksi. Its dielectric constant is 3.75, its index of refraction 1.4585. Its specific heat is 0.18 cal/g-K, its heat conductivity 0.0033 cal/cm-s-K. A very useful property is that its coefficient of linear expansion is only 5.5×10^{-7} per °C. Borosilicate glass (Pyrex) has a coefficient of 33, and ordinary glass 90. Fused quartz will, therefore, withstand thermal stress very well. One indispensable property of quartz is that it is transparent to the ultraviolet, so it can be used for envelopes of ultraviolet lamps for sterilizing, exposing photoresist, erasing programmable read-only memories, exciting fluorescence, and other such duties.

Ordinary glass is called *crown* glass, or *soda-lime* glass, or simply *soft* glass. It is made by fusing sand, sodium carbonate and lime. An average composition by oxides is SiO_2 73, Na_2O 15, CaO 7, MgO 4, Al_2O_3 1. It softens at below red heat, and so is easily worked in the Bunsen flame, as in making chemical apparatus. It even melts in the flame, so it can be *flame-polished.* It is called crown glass not because it is used in crowns, but from the early method of manufacture, which came down from Roman times. A large globe of hot, soft glass is blown, and then the closed end is cut off with glass scissors. In this state it looks like a crown. It is then rotated rapidly to flatten it, and cooled as a round plate. Window panes are cut from it. The "bull's eye" piece where the blowpipe is broken off is less desirable as a window, but makes a good decoration and is even simulated these days. Elizabethan window panes were small and set in lead mullions to make a large window. Large, flat panes were not available until the 19th century.

The *blowpipe* was introduced by the Romans, as well as most of the associated techiques for the hand manufacture of glass objects. Blown glass is not only easy to make, but gives far wider scope to artistic expression than earlier methods, such as casting. A blowpipe is an iron pipe about four feet long, with a mouthpiece at one end. The "gaffer" collects a "gather" from the pot of molten glass in the furnace, and then shapes it by blowing and rotating, and by the use of iron tools. The work is reheated at the furnace to keep it plastic. The Greco-Romans were known for cement and building techniques, but an equally great contribution was the technology of glass.

In *borosilicate* glasses, boron trioxide, B_2O_3 replaces all the soda and lime. This reduces the coefficient of thermal expansion by a factor of 3, so the glass will resist thermal stress, and also increases its resistance to chemical attack. In most parameters, it resembles fused silica, except that it softens at 700°C, a red heat, making it much easier to manage than fused silica, which requires a thousand degrees more.

Flint glasses were developed originally for decorative purposes, to have indices of refraction greater than the usual 1.5 for crown and borosilicate glasses. The name refers to the original use of flints to supply the silica, but this is not an essential matter, and flints were used in making other kinds of glass as well. In flint glass, lead oxide, PbO replaces the lime, and potassium carbonate K_2CO_3 some of the soda. A high-lead glass might be SiO_2 70, PbO 12, Na_2O 10, K_2O 6, CaO 2. The problem is keeping the lead in solution. An index of refraction of over 1.7 can be attained.

Flint glass has less dispersion relative to its index of refraction than crown glass. Dispersion is the change in refractive index with wavelength. For all glasses, the index increases at shorter wavelengths. In lenses, this causes *chromatic aberration*, which was very troublesome in early telescopes with simple lenses. If a strong positive lens of crown glass is followed by a weak negative lens of flint glass, the net power may be positive, while the chromatic differences are equal and opposite and cancel each other. Such a lens is called an *achromat*. Optical glasses have been intensively developed and improved. For an introduction to this subject, see Jenkins and White.

Colored glass is not only very attractive, but has uses as signal lenses as well. Clear glass can be colored by transition metal ions in solution, which are well-known chromophores, or by colloidal suspensions. The colloid particles must be smaller than a wavelength of light, or they would make the glass cloudy. Transition metal ions generally come from dissolved oxides. The principal metals used are Ni, Co, Fe, Mn and Cr.

The colors produced depend on the oxidation state of the metal ion, and on the composition of the glass. Cobalt gives a deep blue, which is much prized. Fe^{++} gives a green colour, but Fe^{+++} makes yellow. Adding FeO to glass gives a green colour, and makes the glass strongly IR-absorbing. Such glass can be used as "heat filters" in slide projectors. Iron is a common impurity in sand, so inexpensive "bottle glasses" may be greenish. The glass insulators used on American telegraph and power lines were of greenish glass. Glass, incidentally, does not make a good electrical insulator because of its propensity of picking up a surface film of water that is conductive.

There are several ways to make a red glass. Colloidal suspensions of selenium or gold may be used. Cuprous oxide reacts to form Cu^{++} ion and colloidal Cu, producing red by two mechanisms. Another source says that the copper must be in the Cu^{+} state to give red; cupric copper gives the usual green. Selenium works better in potash glass than in soda glass. A little selenium and cadmium sulphides added to molten glass give a good result.

Colloidal gold in glass gives a deep red colour. This ruby glass was first made by Kunkel in 1679, when he added Purple of Cassius (a gold precipitate on gelatinous stannic hydroxide, a sensitive analytical test for gold) to molten glass. It can also be made by adding $AuCl_3$ to the glass; the heat is enough to reduce the gold. The same thing happens in aqueous solutions. As little as 0.1% of gold gives a good colour. A little stannous oxide will prevent the coagulation of the gold particles. If the particles become larger, the colour turns to blue.

In colored glass, the colour seen by transmission is the complement of the colour that is absorbed by the ion or scattered by the colloid. If all the shorter wavelengths are absorbed, then the colour will be red. If the green is absorbed, the result is purple. The energy absorbed by a transition ion is lost to the lattice by emission of phonons or other non-radiative process. Gustav Mie became interested in the explanation of the red colour of colloidal suspensions of gold in 1908, and this led to his famous theory of scattering by small particles. Gold particles are hydrophobic (so they are not hydrated). The sol formed by particles 0.04 μm in diameter has a peak absorption at 530 nm, and is red by transmitted light. There is a little yellow-green scattering, but mainly an absorption in the green and shorter wavelengths, with good transparency in the red. A diameter of 0.14 μm gives the *blue* gold sol in which scattering is large in the red. The optical properties of the gold are important in the effect, not just the size of the particles. The effect does not depend greatly on the transparent medium, whether water or glass.

Opaque or translucent glass that is used for decoration may be colored by other means as well. Borax is well known for reacting with metal oxides in glass to produce good colors. Glasses can be made opaque and white by adding oxides that give suspensions of materials of a different index of refraction. SnO_2, TiO_2, CaF_2, As_2O_3, $Ca_3(PO_4)_2$ and cryolite are typical additions. Opaque glass applied to a metal or other substrate by fusion is known as *enamel*, and is an important artistic medium. As was mentioned above, pottery glaze is a glass. The mixture applied to the substrate to be fused into enamel is called a *frit*. This enamel should not be confused with the paint of the same name, which imitates it with an organic film.

Glass, pottery and gems show that there is much beauty in silicon, as well as the everyday utility of semiconductors, stone and cement. The study of silicon leads into many areas of pure science, including the structure and dynamics of the earth, life in the seas and on land, semiconductors, colloids and other areas that we have touched on in this article.

Chapter 7

Boron

The preparation of an article on mercury some time back awakened an interest and curiosity about other metals, so as much information on lead, tin, zinc, copper, aluminium, silver and gold as I could find was assembled into other articles, many of which illustrated interesting chemical or physical subjects. When I happened across boron in the course of these efforts, I realized that I had little familiarity with this element and its compounds, could make little sense of its chemistry, and that boron offered some very interesting lessons. This article is the result. The reader will become much more familiar with the uses of boron, its curious electron-starved chemistry with half-bonds and icosahedral structures, and its similarity to silicon.

THE ELEMENTS

For two thousand years, "the elements" meant earth, air, fire and water, the four Empedoclean elements recognized by Aristotle. Empedocles was a Sicilian who lived 495-430 BC. Influenced by Pythagorean ideas, he taught the permanence of the four elements, and that their mixing in various proportions produced observed matter. These elements are related to what we now call the states of matter: solid, liquid and gaseous, with perhaps plasma added in, but were distinct from them.

This was by no means the only view of nature. The basis of Epicurean philosophy was the atomism of Democritus and Leucippus, which taught that matter was the concourse of infinitesimally small atoms that by their chance combinations created the visible universe, and that physical and chemical changes were caused by the motion of atoms, not the will of spirits. The types of atoms were not specifically indentified, but had to be sufficient to create the observed variety in things. Epicurus died about 270 BC, and the two views existed side by side until the followers of Aristotle and Christianity suppressed and maligned Epicurean views.

Our present views are much closer to those of Epicurus than to those of Aristotle. There was no way to decide between them in antiquity, but the modern atomic theory comes firmly down on the side of Epicururus. Quantitative experiments of the 18th century gave rise to the modern view of the elements, first clearly enunciated by Lavoisier at the end of the century. Now we had oxygen, hydrogen, mercury, carbon and other substances unresolvable into simpler ones, soon multiplied by the electrochemistry of Davy into an extensive stockpile of simple substances.

However, the view persisted of substances with certain properties, like the old Empedoclean elements, but greater in variety and existing as atoms. Atomism explained the law of constant and multiple proportions, and the properties of ideal gases. This is still the naive view of chemical elements. Indeed, chemistry texts still define a molecule as the smallest amount of any substance that retains the properties of that substance. Carbon is thought of as black, silver as shiny, sulphur as yellow, and mercury as liquid, and these are considered properties of the elements.

It is confusing, then, when the combination of two elements does not produce a sort of mean between them, like the offspring of parents. The active soft white metal sodium that burns in air combines with the heavy, pungent green gas chlorine with strong bleaching powers to produce salt, an innocuous transparent brittle crystalline material with a characteristic taste, and which neither burns nor bleaches in any degree.

Elements, in fact, are all pretty much alike. Their atoms consist of a heavy, positively-charged nucleus held together by strong nuclear forces, that holds an equal charge of light, mobile electrons tightly around it, so the whole is electrically neutral and does not exert strong forces on other atoms. Neutral atoms are all roughly the same size, from hydrogen with one proton and one electron, to uranium with 92 protons and an equal number of electrons. Neon (Z = 10) is 0.320 nm in diameter, xenon (Z = 54) is 0.436 nm. Magnesium (Z = 12) is 0.320 nm in diameter, mercury (Z = 80) is 0.300 nm. Individually, they have no macroscopic properties, certainly not those of ponderable quantities of the elements. Atoms are only the building bricks out of which substances are constructed.

When these generic atoms come together, it is often possible to rearrange the electrons, and always the outermost, least tightly held, ones so that the resulting electrostatic energy is less than that of the two separate atoms. How the atoms behave chemically is determined by the possibilities for the redistribution of charge. A little energy is required to extract an electron from a sodium atom, a little recovered when the electron sticks to a chlorine atom, and a great deal of energy is released when a lot of these ions associate in a crystal lattice. Salt is this lattice and its properties, not the properties of any molecule or atom by itself. The textbook definition of molecule is devoid of meaning.

The properties of substances are the properties of associations of atoms, and the nature of the association is more important than the particular atom. Metals are a notable example of this. Most metals are almost alike in nature: soft, shiny, heavy, nontransparent substances that can be bent and hammered. They are a pudding of positive ions in a sea of electrons, and their gross properties do not depend on the nature of the ion. Any suitable atom can make a metal. Sulphur does not do this because other sorts of association give lower energies than a metallic lattice would. Tin exists in both worlds: at higher temperatures it is a typical metal, while at low temperatures it is a crumbly nonmetallic crystalline substance.

I have spoken for simplicity of minimum energy as determining the equilibrium state. At a given temperature, it is actually the minimum free energy F = U - TS that counts, and the entropy S is just as important as the internal energy U. Oxygen molecules attract one another, so minimum energy results when they are all in a pile in the corner. Entropy is what keeps them flying around at room temperature, even at the expense of extra kinetic energy.

Boron will give us an excellent example of these concepts. The generic atom can make substances of the greatest variety and interest. It will be quite clear that there is no such thing as "boron-ness" in the atoms.

Introduction to Boron

Boron is one of the simplest of atoms. The only simpler ones are hydrogen, helium, lithium and beryllium. Boron has chemical symbol B, atomic number 5, and occurs naturally as 80% B^{11} and 20% B^{10}. The latter isotope has a high cross section for thermal neutron absorption, 3800 barns. Thermal neutron counters are often filled with BF_3 gas. The gamma ray from the neutron capture reaction $B^{10}(n,-)B^{11*}$ followed by decay of the B^{11*} to an α plus Li^7 produces ionization which is then detected. Boron is also used in reactor control rods. This is a *nuclear* property of boron, and has nothing at all to do with its chemistry. The atomic weight of boron is 10.81.

Boron is found in a variety of similar minerals all related to *borax*, sodium tetraborate, $Na_2B_4O_7 \cdot 10H_2O$. The name comes from the Arabic *buraq*, "white." Borax is the same in French and German as in English, but the element is *bor*. In Spanish, the words are *boraxo* and *boro*. It is a relatively rare element in the earth's crust, representing only 0.001%.

In the United States, borax is found in large amounts in California, in Searles Lake brines and in the Mojave desert. It is also found in Turkey, South America and other places. The natural deposits are dried-up lake beds. Molten borax reacts with metal oxides to form borates that dissolve in the melt, so it is a useful as a welding and soldering flux, and in colored enamels for iron. In fact, this was the earliest use of borax, as a pottery glaze. This same property is used for borax bead tests in chemistry, where the characteristic colors produced in a transparent borax drop melted on a loop of platinum wire in a bunsen burner flame are observed. Blue, for example, is the colour of cobalt; green, of chromium. The colour can differ in oxidizing (blue) and reducing (yellow) flames.

Borax hydrolyzes in water to form a slightly alkaline solution that is good for cleaning, since it emulsifies grease and oil. It also softens water by precipitating calcium borate. A household hint for preparing borax soap for really difficult laundry jobs was as follows. Cut up one pound laundry soap, add to one ounce of borax and a quart of water, and boil until uniform. Allow to cool, apparently in a mold, and cut into bars when solid.

The electron configuration of boron is $1s^22s^22p$. It has only three electrons to work with, so the ion is unpolarizable, and does not hydrate. For this reason, boron is not eager to donate electrons in an electrovalent bond, and can also not accept them easily. Therefore, most of its bonds are covalent, and even forms *half-bonds* in which only one electron is shared covalently, not the usual two. This gives boron an apparent valence of +6 that we shall see in some interesting compounds. The first ionization potential is 8.30 V, which is not unusually high.

The other commonly-used boron compounds are orthoboric acid, or simply boric or boracic acid, H_3BO_3, and boron trioxide, B_2O_3, its anhydride. Note that the formula for boric acid can be written $B(OH)_3$, as boron hydroxide. If boron were a normal metal, the hydroxyl ions would separate in water, creating the trivalent boron ion B^{+++}. This, however, does not happen to the smallest degree, and boron does not form ionic bonds. Moreover, boric acid is not gelatinous like aluminium hydroxide, but crystallizes nicely. Boron trifluoride, BF_3 is not an ionic compound like NaF, which has no ions in its crystals. Instead, the hydrogens are lost long before the oxygens. In aqueous solution, boric acid is a very weak acid, weaker even than carbonic. Its first ionization product is 6.4×10^{-10}. Boric acid solution (solubility 6.35% at room temperature, 27.6% at 100°C) is used as an antiseptic, especially as an eyewash. Boric acid can be produced from borax by treatment with H_2SO_4, and boron trioxide by heating boric acid. These are the initial processes in the synthesis of all boron compounds.

Elemental boron was discovered by Davy, Gay-Lussac and Thınard in 1808. They produced metallic potassium by electrolysis, and then used it to reduce borates to impure boron. Davy called the element *boracium*, the Frenchmen *bore*. It can also be obtained in impure form by reduction of the oxide B_2O_3 by magnesium, or in pure form by the reduction of BCl_3 by hydrogen on hot filaments. The first pure boron was produced by Weintraub in 1909. Ordinary boron is a brown-black amorphous powder. Pure boron can be made as extremely hard yellow monoclinic crystals that are a semiconductor resembling silicon. The band gap is 1.50 or 1.56 eV. Crystalline boron is an insulator at low temperatures, but becomes a conductor at elevated temperatures, as would be expected as carriers are thermally excited into the conduction band. Fabrication difficulties have so far prevented the use of boron as a semiconductor. The density of crystalline boron is 2.34 g/cc, of amorphous boron, 2.37. It melts at 2300°C and boils at 2550°C (some sources say 2040°C and 4100°C), so it is a very refractory substance. Boron fibers have been used in composite materials because of their great strength.

Boron gives a blue-green flame, and the brown amorphous form is often used in pyrotechical devices for this purpose. The colour can be distinguished from the emerald green of copper. Boron is also used in borosilicate glasses, which are 12%-15% B_2O_3, 80% SiO_2, and 2% Al_2O_3. Sodium, potassium, magnesium and calcium oxides are kept to minimal amounts, since it is their exclusion that gives the glass its desirable properties. "Pyrex" is a common trade name for a borosilicate glass. This glass is chemically resistant, and has a small coefficient of thermal expansion. Boron carbide, B_4C, harder than SiC, is formed by decomposing B_2O_3 with carbon in the electric furnace: $2B_2O_3 + 7C \rightarrow B_4C + 6CO$. It was first suggested for commercial use in 1934, and is an excellent abrasive. Boron nitride, BN, is another very hard compound, used in cutting tools. There are numerous other borides with complex structures. Boron is also used in porcelain enamels for iron, and for tiles and sanitary ware.

An interesting test for boron and borates uses volatile $(CH_3)_3BO_3$, which burns with a green flame. Combine the solution to be tested with methanol, then add concentrated sulphuric acid to make the methyl borate. Another test uses turmeric paper, which turns reddish brown when moistened with boric acid or borates. If then a drop of ammonium hydroxide or sodium hydroxide is put on the paper, it will turn dark green, blue or black. To make turmeric paper, wash some ground turmeric in water. Then digest the washed powder with alcohol, and filter. Soak unsized white paper in the filtrate, then dry and cut into test strips. The active ingredient in turmeric paper may be curcumin, which seems to act the same way. It is available in an alcoholic 0.1% solution for this purpose.

Borax is necessary in small amounts for plant growth, one of the 16 essential nutrients. In larger amounts it is poisonous to plants, and the range can be small. For peaches, 1 ppm is required, but more than 5 ppm is toxic. If the signs of boron deficiency are noted in plants, a boron supplement can be applied. Borates can be used as non-toxic and non-specific herbicides. Borates are non-toxic to animals. The LD_{50} (dose at which there is 50% mortality) for humans is about 6 g per kg of body weight. Anything above 2 g is considered non-toxic, and borates are only 2 to 3 times as toxic as aspirin. Therefore, you are pretty safe unless you eat a pound and a half of borax for a snack. Borates are more toxic to insects than to mammals. The boranes and similar gaseous compounds are quite poisonous. As usual, it is not an element that is intrinsically poisonous, but toxicity depends on structure.

Boron-Oxygen Chains

We have mentioned that boron would rather lose the H in $B(OH)_3$ than the oxygen. Like silicon, which has a similar persuasion, boron can form chains like -B-O-B-O-B-, where each intermediate boron has a free valence to work with. This is seen in tetraboric acid, O=B-O-B(OH)-O-B(OH)-O-B=O, or $H_2B_4O_7$, which can be derived from $B(OH)_3$ by dehydration: $4H_3(BO)_3 \rightarrow H_2B_4O_7 + 5H_2O$. Dehydration of boric acid also gives boron trioxide: $2B(OH)_3 \rightarrow B_2O_3 + 3H_2O$. Boron trioxide can also be written O=B-O-B=O, where the boron-oxygen chain again appears. Some compounds can be considered derivatives of the theoretical acid HBO_2, called *borates*. This metaboric acid is boric acid less one water molecule. If it loses another water, it becomes B_2O_3, which we have already seen. Boron trioxide in water becomes boric acid. The use of borax as a flux depends on reactions like O=B-O-B(ONa)-O-B(ONa)-O-B=O (borax) + NiO $\rightarrow$ $2NaBO_2$ + $Ni(BO_2)_2$.

Boron-oxygen chains also form linked planar structures that have the appearance of fence wire with hexagonal openings. Oxygen atoms project radially to make bonds with other atoms.

Icosahedral Boron

The icosahedron ("twenty-base") is one of the five Platonic solids, a regular convex polyhedron with 20 faces, 30 edges and 12 corners. In the absence of degeneracy, a symmetrical structure will have the lowest energy, and the regularity is possible because of the identity of the atoms. The tetrahedron, square, octahedron and dodecahedron also appear in molecular structures. It is remarkable that boron finds the icosahedron adapted for its purposes. Excellent drawings of the structures we discuss here are found in Pauling and Hayward. The icosahedral structures are exhibited by compounds known as *boranes*.

Just after world war II, the U. S. military was interested in developing an advanced aviation fuel for jets and ramjets that would replace the hydrocarbon-based JP-4 and similar fuels. Someone suggested that the then poorly-known boranes would give a better power-to-weight ratio, and Project ZIP was launched in 1952. Government money rained down on research projects dealing with boron, and for roughly a decade there was near hysteria in the scramble for research money, and many things were, incidentally, learned about boron. The boranes turned out to be not the powerhouses that were anticipated, were hard to manufacture and store, tended to start fires, and were also extremely toxic. Military enthusiasm cooled, and the research money went elsewhere. However, we found out the interesting structures of boranes, and a good deal of boron chemistry. Fifty years later, airplanes still burn JP-4.

A boron atom at an icosahedral vertex can make five equal bonds to its nearest neighbors, leaving one bond to stick out radially. Boron-boron bonds are half-bonds of length 0.180 nm. If we put one boron atom at each vertex, each with one proton, and assume that all bonds are half-bonds, then we have the neutral structure $B_{12}H_{12}$. However, the protons would like to have more electrons in their bonds, and overall 6 electrons are lacking for a full single bond. This would give far too large a charge to the ion, so the structure only attracts 2 extra electrons, which are spread over the 12 bonds to the protons. It happens that two extra electrons are ideal for resonance stablization of the ion. The proton bonds wind up at 0.120 nm. The substance potassium dodecaborohydride contains this ion in its crystals.

Boron will not form BH_3. Instead, it forms larger borohydrides called boranes. In decaborane, $B_{10}H_{14}$, only 10 of the 12 vertices of the icosahedron are occupied. The extra 4 hydrogens are each bonded to two adjacent boron atoms, replacing the "missing" ones. The 10 radial hydrogens have bonds 0.118 nm long, while the bridging hydrogens have bonds 0.135 nm long. In tetraborane, B_4H_{10}, only four vertices are occupied, and there are 10 bridging hydrogens. Tetraborane has a foul smell. Diborane, B_2H_6 has no boron-boron bonds, but two bridging hydrogens out of plane in the planar molecule.

Davy might have made a borane when he treated boron with hydrochloric acid. The gas that was evolved burnt with a blue-green flame. Boranes react vigorously with oxygen, so they have been considered as candidates for rocket fuels. They are examples of *electron-deficient* compounds, since there are more bonding orbitals available than there are electrons to fill them. *Carboranes* are boranes where some of the borons are replaced by carbons. It is not easy to make boranes, and there was a considerable interval between Davy and the first recognized boranes.

One form of crystalline boron is tetragonal. The unit cell of this lattice consists of 4 icosahedrons of boron and two extra boron atoms, 50 in all. The radial bonds go to other icosahedrons or to the extra atoms, which show tetrahedral bonds. The bond length is 0.180 nm, as in the boranes.

Mineralogy and Production of Boron

Borax was first known from the deserts of western Tibet, where it received the name of *tincal*, derived from the Sanskrit. Borax glazes were used in China from AD300, and some tincal even reached the West, where the Arabic alchemist Geber seems to mention it in 700. Marco Polo brought some back to Italy in the 13th century. Agricola, around 1600, reports its use as a flux in metallurgy.

In 1777, boric acid was recognized in the hot springs or *soffioni* near Florence, Italy, and became known as *sal sedativum*, with mainly medical uses. The rare mineral is called *sassolite*, which is found at Sasso, Italy. This was the main source of European borax from 1827 to 1872, at which date American sources replaced it.

Borax was found by John A. Veatch in the spring waters of Tehama County, California, in January 1856. By September, borax was being produced from Borax Lake in Lake County. The California Borax Company worked these deposits from 1864 to 1868, when the company failed. John Searles and others discovered borax near Mono Lake in 1870, and this "marsh borax" became the principal source. One of the minerals exploited was *ulexite*, $CaB_4O_7{\cdot}NaBO_{2{\cdot}8H2}O$, or "cotton balls," since it is found in soft, fibrous balls.

It is common in Chile, Argentina, Nevada and California, where it is found in desert playas. The fibers guide light like fibre optics. Searles worked these deposits until 1896. However, in 1882 colemanite was discovered further south in the Mojave desert, and mining began near Daggett in 1887. This source in the Calico Mountains was much more accessible, since it was near the railway. It was worked by underground mining. In 1903, on exhaustion of the Calico Mountain reserves, mining moved north to Death Valley, and Colemanite began to be mined in open pits. Colemanite is a calcium borate formed by alteration of ulexite. It is named after the San Francisco merchant who first marketed it. Before the discovery of colemanite, South America supplied most of the world's borax, which had been discovered in Chile in 1852, but not worked commercially until much later.

The Death Valley colemanite mines were the origin of a famous trade name. The Harmony Borax Works were established at Furnace Creek in 1883. The product was hauled 166 miles south to Mojave by teams of 20 mules. The size of these teams was probably due to the harsh environmental conditions, in order that the mules should not exhaust themselves in the heat and aridity. They pulled only a wagon and a tank trailer (which was probably their water), which normally could probably have been handled by six mules. "20-Mule-Team Borax" was seen on many store shelves for many years, though now it is difficult to find borax in any supermarket. The teams worked until 1890, when the company failed and the mine was closed.

Kernite is an altered form of borax with only 4 waters of hydration instead of 10. It is also called *rasorite*, and has been the principal source of borax since 1927. Another source is the waters of Searles Lake, which are worked for other minerals as well, such as trona. Kernite is only found in Kern County, California, in the Kramer district about 3 miles north of Boron, California. The earlier railway siding at Kramer was renamed Boron. These deposits were discovered in 1916. Colemanite had been produced just to the east after 1913. Kramer kernite and Searles Lake brines have now completely superseded colemanite production. The Kramer lake beds are found in the Pliocene Ricardo formation. Most production is now from an open-pit mine opened in 1955. Boron is about halfway between Barstow and Mojave.

Borax is found mainly in arid regions, since borates are soluble to some degree and in a humid region would have been leached away long ago. They were deposited originally from waters associated with vulcanism. In California, this would be the great Tertiary vulcanism typified by sheet basalts. These borax-laden waters evaporated in lakes without an outlet in the arid environment that existed then as well as now. Some of these deposits were covered by later sediment and protected from leaching, like the kernite deposits of Boron. The conversion from borax to kernite is a result of mild metamorphism caused by burial under several hundred feet of overburden.

Where sea waters have evaporated instead of continental waters, as at Stassfurt, Germany, the hard *boracite*, a magnesium borate and chloride, may be found, associated with halite and gypsum. It is the only hard boron mineral

(H = 7), though colemanite (H = 4) is harder than borax minerals.

Chapter 8

Iron

Iron is best known as the metal that gave us weapons and tools, and whose ability by means of alloys and heat treatment to suit itself to every application makes it the primary metal of technology. The purpose of this article, however, is principally to explain in some detail two remarkable applications of iron, *ferromagnetism* and oxygen transport in the blood. The very similar cobalt and nickel will also be included.

Iron is the most freqently encountered metal in daily life, always in the form of manufactured objects, and usually covered with a protective coating or buried deep within the object. Concrete structures contain essential reinforcing iron; electrical machines, including transformers, depend on iron. Automobiles are mainly iron; "tin" cans are iron covered with thin coatings of tin or lacquer; fasteners, such as the nails and screws used in wooden construction, are usually iron. This list can easily be extended with a little thought.

Iron is an excellent and versatile material of construction—strong, tough, easily formed and worked, and, very importantly, cheap compared to the alternatives. Plastics give it competition, especially in products that must be manufactured at the lowest cost where strength and durability are not the primary concerns, such as modern American automobiles. Aluminium is a strong competitor where weight is a concern, as in aircraft. However, the versatility of iron-carbon alloys cannot be matched in any other material. Alloys with other metals, such as nickel, chromium and manganese, give further advantages. These *steels* can be tailored to nearly every demand, and are not significantly challenged as materials of construction.

The shortcomings of iron are its weight, and its propensity to rust. Both aluminium and plastics avoid these shortcomings, because their densities are much less than that of iron, and aluminium is protected by an adherent layer of oxide, and plastics by the inertness of the substance. The advantages of iron are so great, however, that these considerations prevail only in limited fields of application.

The three metals iron, cobalt and nickel are called the "iron family" and are very similar, so I shall discuss them together. A property of the iron family for which there is no substitute at an equivalent cost is that they can be induced to provide a strong magnetic field with only small excitation by an electric current. It is a very remarkable property, so an effort will be made to explain how it arises, and how it is used. Also, iron is important as a carrier of oxygen in blood, and how it does this will be explained.

In addition to these subjects, the physical and chemical properties of iron will be reviewed, and its curiosities examined, as usual. For some accounts and explanations, reference will be made to other articles where they appear, and will not be duplicated here. This is especially true of the application of iron as an engineering material, a very extensive subject.

Our word *iron* is cognate to the German *Eisen*, which in various forms is found in all the northern European languages, though not elsewhere. Why our pronunciation of the word is metathetical, "iorn" instead of "iron," I have not heard. In Gothic, it was *eisarn*, and in Old High German, *isan*. In Greek, iron is *svderos* and steel is *chalybicσn*, stems that are encountered frequently in talking of iron, as in "siderurgy" and "chalybeate spring." "Sidereal time" is not iron time, however, but the "sider-" comes from the similar Latin *sidera*, "stars." Iron was very valuable when it came to the bronze-age Greeks from its inventors around the Pontus Euxinus, and was used for jewellery and prizes. It has been money in other places and times. The Latin for iron is *ferrum*, from which the word in most European languages has been derived, and which appears more commonly than any other stem in words dealing with iron, especially in chemical nomenclature.

Cobalt is from *Kobold*, an earth spirit, or a good house spirit. The kobold came surreptitiously and stole the silver from the ore, replacing it with base cobalt. There is a Greek word *kobalos*, meaning "rogue, trickster," but there is probably no connection with the German. Nickel is from *Nickel*, a water spirit, who took the copper from the ore and washed it away, replacing it with *kupfernickel.* A *Nix* is a male water spirit, a *Nixe* a female water spirit. These names were from miner's slang, not traditional names for the metals, which were unknown at the time. Many minerals resemble ores, but do not yield the expected metal, and this was confusing when their chemical natures were not known. Cobalt was recognized by Brandt in 1735 and nickel by Cronstadt in 1750, but their compounds were not carefully studied until the next century.

PROPERTIES OF IRON

Iron is a shiny, bright white metal that is soft, malleable, ductile and strong. Its surface is usually discolored by corrosion, since it combines readily with the oxygen of the air in the presence of moisture. In absolutely dry air, it does not rust. The oxide that is produced is crumbly and soft, giving no protection to the base metal, which eventually rusts away. It is found in nature as the metal only in meteorites and in very rare circumstances where iron minerals have been reduced by environmental factors. Masses up to 25 tons in weight have been found in West Greenland. Practically, it is always obtained from ores that are usually the oxides, and occasionally the carbonate, as low in sulphur and phosphorus as possible. The plentiful iron pyrite, FeS_2 is not an acceptable ore because sulphur is a deleterious impurity in iron, and expensive to remove in refining. Iron is the fourth most plentiful element in the earth's crust (4.6%), but is so widely disseminated that it can be obtained only from ores in which iron has been specially concentrated. Much of this concentration, incidentally, occurred very early in the earth's history when iron removed from the atmosphere the little oxygen that it then contained, possibly with the help of the earliest forms of life.

A clean surface of cobalt is tinged with pink, nickel with yellow. Neither rusts as vigorously as iron, since the oxide layers are more protective and adherent. Nickel displays a resistant bright surface that takes a high polish. It is generally used as an undercoat for chromium plating, since chromium is not suitable by itself on iron, as the electrodeposited film is not continuous.

Iron has atomic number 26, atomic weight 55.85, and stable isotopes 54 (5.9%), 56 (91.6%), 57 (2.2%) and 58 (0.33%). Its electron configuration is $Ar3d^6 4s^2$, and its first and second ionization potentials are 7.87V and 16.18V. With its neighbors cobalt (Z=27) and nickel (Z=28), it is one of the "iron triad" of similar metals. It is in the centre of the periodic table, in the region of "transition metals" where a d-shell of electrons is being filled. The 4s electrons are actually more stable than the 3d electrons, so the d-electrons are actually on the outside of the atom. The d-shell can hold 10 electrons, and as it becomes nearly filled, drops below the 4s electrons in energy. All these atoms filling d-shells make metals that are very much alike; if the d-electrons were more inside, these metals would be even more alike than they are. More will be said in connection with magnetism, in which these electrons play leading roles.

At room temperature, iron is in the form of *ferrite*, or α-iron, a body-centered cubic structure. The density of α-iron is 7.86 g/cc. At 910°C it changes to γ-iron, which is face-centered cubic and somewhat softer. At 1535°C iron melts, and boils at 3000°C. For more information on iron structures and the iron-carbon phase diagram. Cobalt melts at 1480°C, nickel at 1455°C. The specific heat of any of the three metals is about 0.107 cal/g-K. The thermal conductivities of Fe, Co and Ni are 3.37, 3.81 and 4.19 cal/s-cm-K. Their electrical resistivities are 9.71, 6.24 and 6.84 μΩ-cm. These are "worse" than those of copper by factors of only 4 to 6, so the iron metals are very good conductors of electricity and heat. Comparing the numbers shows how similar these metals are in their physical properties. I have not heard whether cobalt and nickel make useful alloys with carbon, as iron does. They are much too expensive to use as structural metals, other than as alloying elements or coatings.

Iron occurs as the cations Fe^{++}, ferrous iron, and Fe^{+++}, ferric iron. Nickel and cobalt occur mainly as the dipositive ions, nickelous and cobaltous, but also in other oxidation states, usually in unstable compounds. A large number of important salts are formed with various anions. Fe^{++} will gladly donate an electron in an acid solution or in the presence of oxygen, so it is a reducing agent, oxidizing itself to Fe^{+++}. In alkaline solution, this ion will gladly accept an electron, so it is an oxidizing agent, reducing itself to the ferrous state. In acid solution, Fe will reduce H_+ to H_2, since its electrode potential is -0.44V, well above hydrogen's. Iron is, in fact, an active element.

The oxides are ferrous oxide, FeO and ferric oxide, Fe_2O_3. Ferrous oxide is not stable against partial oxidation to $Fe(II)(Fe(III)_2O_4)$, usually written Fe_3O_4 or $FeO{\cdot}Fe_2O_3$. This hard, black substance is *magnetite*, an important ore of iron and a very interesting substance in its own right, which will be considered further in connection with magnetism, where its structure will be explained.

The ferrous ion is greenish in solution, while the ferric ion is a light violet. The distinct colors of iron compounds are due to the d-electrons, which can interact with light in many interesting ways. Ferrous oxide, as a mineral, is called *hematite* ("blood-stone") and is usually almost black. It leaves a red streak on unglazed porcelain, and in its usually finely-divided condition is a characteristic bright earthy red. In pure form, it is called *rouge*, good for reddening cheeks and polishing glass, or *Venetian red*, a red pigment. The red stains on concrete are ferric oxide, usually from reinforcement rods that are not properly protected from the weather. The red of Mars is ferric oxide, showing that the atmosphere once contained oxygen, the oxygen that is only liberated by life.

Ferrous sulphate, $FeSO_4$ forms pale green crystals of the

heptahydrate when a solution is evaporated. In air, these crystals will attract moisture and oxygen, and slowly become ferric. This salt is called *copperas* or *green vitriol.* In alkaline solution, gelatinous $Fe(OH)_2$ precipitates, which likewise oxidizes in air through green, black and finally to red-brown ferric oxide. Ferrous tannate is obtained by mixing ferrous sulphate and tannic acid (from nut-galls, for example). This oxidizes in air to black ferric tannate, and so the solution is used as an ink. A dark-blue dye is generally added, since the ferrous tannate is almost colorless. By the time the dye has bleached in the light, the ferric tannate will be distinct. This kind of ink eventually rusts to ferric oxide, and is not nearly as permanent as India ink made from colloidal carbon. It is not as susceptible to clogging, however, and flows easily from the pen. It can be removed from fabrics with oxalic acid or ammonium oxalate, which dissove the iron. This was called "permanent ink" to distinguish it from "washable ink" that contained water-soluble dyes and was even less durable.

Anhydrous ferric chloride is a covalent compound $FeCl_3$ that dissolves in organic solvents as well as in water. It forms bright-yellow crystals as the hexahydrate. Anhydrous ferric sulphate is colorless, but crystals of $Fe_2(SO_4)_3{\cdot}9h_2O$ are yellow. Ferric ammonium alum, $FeNH_4(SO_4)_2{\cdot}12H_2O$ is pale violet, the typical ferric colour in solution. This is an "alum" without aluminium, the iron standing it for it. It is a mordant because of the gelatinous ferric hydroxide formed when a solution is make alkaline. There are also ferric potassium alum, also colorless or violet, and ferrous ammonium sulphate, which forms blue-green crystals of the hexahydrate.

The cyanide ion, CN^-, and its relatives, accompany iron in many interesting compounds. The free cyanide ion is very poisonous, since it blocks nerve communications. This poison is purely in the structure, again, and there is nothing inherently poisonous about cyanide. Iron forms strong complexes with six cyanides, binding the carbon tightly and pointing the nitrogens to the vertices of an octahedron. If the iron is ferrous, the charge is -4, and if it is ferric, the charge is -3. This is the *ferrocyanide* or *ferricyanide* anion, which makes many salts in aqueous solution or ionic crystals. It even combines with iron, to form ferric ferrocyanide, ferric ferricyanide, ferrous ferrocyanide and ferrous ferricyanide, and often potassium joins in. Potassium ferrous ferricyanide is *Prussian blue*, ferrous ferricyanide is *Turnbull's blue*. Potassium ferrocyanide, also called potassium berlinate or potassium prussiate, is yellow, and can be obtained in the distillation of coal. Potassium ferricyanide is red. Solutions of ferrocyanide and ferricyanide show no evidence of either iron or cyanide, and are not poisonous. An Italian terrorist who plotted to poison with ferrocyanide was doomed to failure by his ignorance. Some colorless ferricyanides turn the urine blue, which was a typical prank among chemistry students. Prussian blue is now marketed as the prescription medicine Radiogardase, which aids elimination of thallium and caesium, beneficial when radioactive isotopes of these have been ingested, and also in cases of thallium poisoning, which is extremely serious. Caesium does not appear to be very poisonous. The new name would seem to denote an enzyme, which it is not, but at any rate allows sale at a much higher price.

The cyanate ion, CNO^- and the thiocyante ion, CNS^- are equally stable and nonpoisonous. Ferric thiocyanate, $Fe(CNS)_3$, is blood-red and an analytical test for iron. It forms red, deliquescent crystals. When carbon monoxide is passed over warm Fe, Co or Ni powder, liquid *carbonyls* are formed, $Fe(CO)_5$, $Co(CO)_4$ and $Ni(CO)_4$ the uncharged CO sticks to the Fe or Ni by "resonance" with states where various formal charges are assigned to the atoms, and a linear combination leads to a minimum energy. The carbons are adjacent to the central metal. A mixture of carbonyls can be fractionally distilled and the pure substances pyrolyzed to recover the metals. The Mond process for refining nickel is based on this property.

MINERALOGY AND NATURAL HISTORY OF IRON

The usual ferrous sulphide, FeS, can add sulphur stochiometrically to form crystalline FeS_2, known as the mineral *pyrite*. This is from the Greek *pyrites*, from *pyr*, "fire," and meaning something made of fire. When struck by iron, it will produce luminous sparks that can be used to ignite tinder to make fire. The *wheellock* was invented around 1500 for use in cavalry pistols. A spring-driven wheel pressed to a piece of pyrite threw a jet of sparks into the pan. Cavalrymen could not use the usual "slow match" of the time because the wind extinguished it. Pyrite gives excellent sparks, probably pieces of FeS_2 heated by impact and burning in air, but is not as durable as flint. Such materials may be called *spintharogenic*. Good crystals of pyrite are an excellent brassy yellow colour, leading fools to mistake it for gold. However, it is hard and light, not soft and heavy, so it is easily distinguished from gold. In small flakes, however, it could be deceiving, and especially because it is very often associated with gold.

Pyrite crystallizes in the cubic system, and its crystals are typically cubes and pyritohedrons. A pyritohedron has 12 pentagonal faces, but the pentagons are not regular, so the solid is not a dodecahedron. Crystals do not have 5-fold axes. The cube faces are usually striated, each face at right angles to the neighboring faces. Hardness, 6-6.5, and density 4.8-5.1 g/cc. When heated, sulphur is driven off and the iron oxidized to magnetite. Marcasite is orthorhombic instead of cubic, but the same substance. Marcasite is a whiter gold colour. Pyrrhotite, a third sulphide, is closer to FeS. It is metallic bronze or brown-colored, softer (H = 3.5-4.5) and a bit lighter (4.40-4.65 g/cc) than pyrite. Notably, it is magnetic, probably ferrimagnetic. Its crystals are hexagonal. Pyrrhotite often contains nickel (as at Sudbury, Ontario) and is one of the most important ores of that metal.

Pyrite is a good source of sulphur, but not a good source of iron because of the sulphur, which ruins iron and is hard to separate from it. It has been used for the manufacture of sulphuric acid. Copperas, the sulphate, is obtained by its oxidation. Mineral copperas often accompanies pyrite as a weathering product. The "Sparta dollars" of southern Illinois are thin, round, radial pyrite crystals deposited in the bedding planes of shale. Pyrite is also found in coal, where it adds unwelcome sulphur to the fuel.

Magnetite may be the most important ore of iron. It is found at Kiruna, Sweden, and in other regions of ancient and metamorphic rocks. It (rarely) forms beautiful shiny black octahedral crystals, as well as rhombododecahedrons. It is hard, 5.5-6/5 and rather heavy, d = 5.18 g/cc. It dissolves in hydrochloric acid, and, of course, is strongly magnetic. Lodestone is a magnetized lump of magnetite that attracts iron. Lodestone is found at Magnet Cove, Arkansas, among other places.

Hematite and limonite are ferric oxide, Fe_2O_3. In massive form, hematite is black, but when finely divided, red. There is an earthy variety of hematite that is brick red and a good paint pigment. It was called *reddle* and used to paint distinguishing marks on sheep. *Reniform* hematite looks like beef kidney, with metallic lustre. It is usually the product of oxidation of other iron minerals, often with secondary enrichment. The exhausted rich hematite ores of the iron ranges of Minnesota, Wisconsin and Michigan were formed from iron silicates and and carbonates of early sedimentary origin. The Clinton iron bed of Alabama is also secondary hematite, as are the Brazilian ores.

Limonite is colloidal hydrated ferric oxide, yellow to brown in colour. It is light, density 3.6-4.0 g/cc. Used as paint, limonite is known as *ochre* (see below). It is widely distributed, as in the Jurassic ores of eastern England (no longer exploited). It is a supergene concentration of iron from sedimentary rocks. Colloidal materials were long called amorphous, but they are actually microcrystalline. Goethite is crystalline hydrated ferric oxide, FeO(OH), much like limonite but crystalline (orthorhombic). It is brownish-black, yellowish or reddish. It was found in Cornwall, Alsace-Lorraine (the "minette" ores), Altenberg in Saxony, Lake Onega in Russia, and Jackson Iron mine, Michigan. Goethite was named in honour of the poet and scientist J. W. von Gφthe (1749-1832).

Bog iron ore is limonite or goethite, deposited with the aid of *iron bacteria* in anaerobic swamps. The iron is carried in by waters containing CO_2 as soluble bicarbonates. The organic matter reduces it to oxides, which make characteristic iridescent patches on the water, then precipitate on the bottom of the marsh to form an iron-rich layer. It was smelted at Lynn, Massachusetts with oyster shells and charcoal to be hammered into bars to be traded for slaves in Africa, which were sold in the Caribbean for sugar to be brought to New England for fermentation and distillation into rum to be sold to the Indians, making the fortunes of Boston and Providence. The Lynn ironworks have been reconstructed, but with questionable fidelity.

Ferric oxide is the basis for many pigments of earthy colors, generally through forms of the minerals hematite and limonite, but also through the weathering of sandstones containing an iron cement, or of ferruginous shales. The most familiar colour is *ochre*, also known as raw sienna or oxide yellow. To me, it is a tan, not a bright yellow like cadmium yellow. Burnt sienna is darker and redder, perhaps a tuscan red. Raw umber is definitely brown, and burnt umber a dark brown that is called Vandyke brown. The colour names are drawn from the ferruginous earths of Siena and Umbria. Ingredients other than ferric oxide, such as the manganese in sienna earth, may give characteristic shadings of colour. Colors are notoriously difficult to name and describe, so a good colour book is a great help.

The iron group are the only metals that fall from the sky. There is a lot of dıbris in space that enters the earth's atmosphere at 10-30 km/s relative velocity (much of this is the earth's orbital velocity of 30 km/s). These particles, mostly microscopic, cause the trail of a *meteor*. A lot of these are cometary particles, ices and dust, but these never reach the ground, burning up between 50 and 150 km altitude, roughly the domain of the ionosphere. A bright magnitude-0 meteor is less than 8 mm in diameter. Millions of meteors burn up in the atmosphere every day, mostly unseen. I make it that they add about 15 kg a year to the earth, but an astronomy text says 10,000 tons a year, which I believe is only enthusiasm and a missed decimal point. The odd large meteorite could put a spike in the average, of course. The larger lumps, of which there are few, reach the ground before burning or blowing up, and are called *falls*, or meteorites. Only Mrs. Hodges of Sylacauga, Alabama has been injured by one, on 30 November 1954. 92% of the falls are *stony* meteorites, 6% *iron* meterorites, and 2% *stony-irons*. The stony meteorites are mostly *chondrites*, which contain characteristic small glassy spheres. The rest are *carbonaceous chondrites*, containing volatile matter, and a few *achondrites* that seem to have been heated to a high temperature. The composition of stony meteorites is very much like that of olivine, 36 O, 23 Fe, 18 Si and 14 Mg. The remaining 9% is very miscellaneous. The average composition of olivine is 38 O, 28 Fe, 17 Si and 18 Mg.

Iron meteorites are nearly pure iron-group metals: 90.7 Fe, 8.5 Ni, 0.6 Co is an average composition. The remaining 0.2% is negligible. They are the most often picked up, since they have a distinctive appearance, unlike the stonies, which look just like stones. When sliced, polished and etched they show the Widmanstδtten pattern that shows they cooled slowly. In studying them, Widmanstδtten originated metallurgical microscopy. Meteorites are not cometary rubbish; astronomers say they look like the fragments of planetoids like asteroids. The composition of meteorites suggests the composition of the earth: a mantle of olivine covering a ball of iron group metals. The volume ratio of mantle:core is 82:17, not too far from the ratio of stony:iron falls of 92:6.

Iron and Magnetism

The magnetism of the iron group of metals is a rare and remarkable property. It is not due to any inherent magnetic propensities of the atoms, but to the *structure* of the metal. Other substances with similar structures also have similar magnetic properties. The properties that we will explain are called *ferromagnetism*, *antiferromagnetism* and *ferrimagnetism*. All three are basically similar, but have different external expressions.

A magnetic field H is created by electric currents. It exerts forces on a moving charge q given by $F = (q/c)v \times H$. Here the charge q is in esu, c is the speed of light, about 3×10^{10} cm/s, and the magnitude of the magnetic field H is in oersted, which is numerically equal to the magnetic flux density in gauss. In space, there is no difference between field and flux density. Wherever there is a magnetic field, there must be moving charge somewhere to produce it.

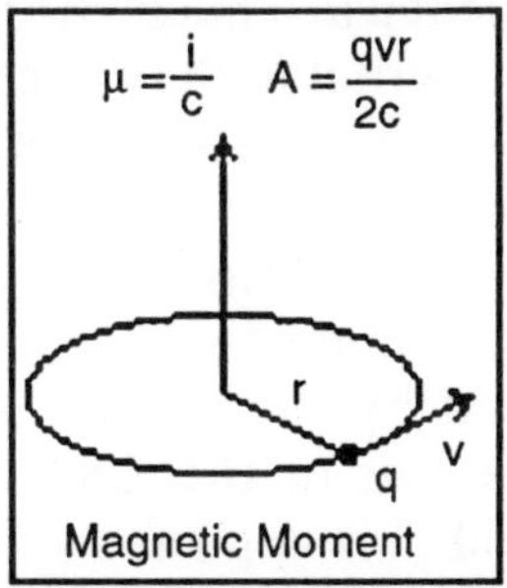

Magnetic Moment

An electron moving as in a Bohr atom around a nucleus represents such a current, which produces a magnetic field. If q is the charge, and v is its orbital velocity, then the average current in an orbit of radius r is $qv/2\pi rc$. The product of the current and the area of its orbit is called the *magnetic moment* μ of the current. Hence, $\mu = qvr/2c = q(mvr)/2mc = (q/2mc)(mvr) = (q/2mc)j$, where j is the *angular momentum* of the mass m. This is quite a general result, and the quantity in parentheses is called the *gyromagnetic ratio*. The *orbital motion* of the electron causes a magnetic moment $\mu = -(e/2mc)j$ antiparallel to the angular momentum.

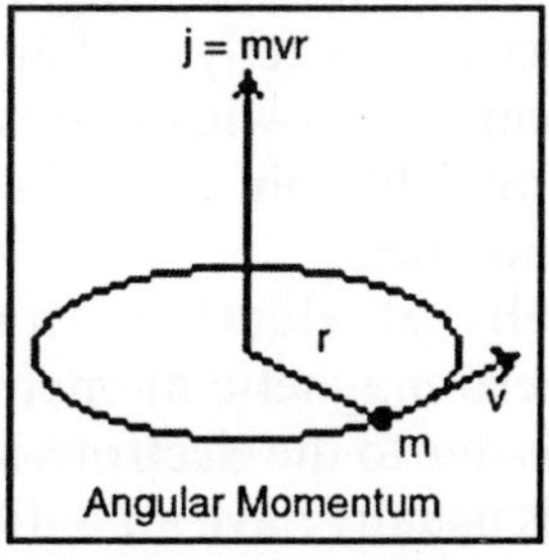

Angular Momentum

The deBroglie wavelength associated with a momentum p is $\lambda = p/h$, where the dimensions of Planck's Constant h work out to erg-sec. The circumference of an orbit of radius r is

$2\pi r$, so if an integral number of wavelengths is to fit into it, $2\pi r/\lambda = n$, an integer. This demands that the angular momentum be an integral multiple of $h/2\pi$: $j = n(h/2\pi)$. This means that the magnetic moment will be quantized in multiples of $eh/4\pi mc$, which is called the Bohr magneton, μ_B, about 0.927×10^{-20} emu.

A magnetic moment μ in a magnetic field H is acted upon by a torque tending to turn it in the direction of the field. The energy function that yields this torque on differentiation with respect to the angle is $U = -\mu \cdot H$. The energy differences resulting for various orientations of the atomic magnetic moments should appear in measurements of the wavelengths emitted and absorbed by the atoms, and indeed it does, especially in observations of the Zeeman Effect, when an external magnetic field H is applied.

These measurements (and others) revealed a surprising fact: the electron itself must be regarded as possessing a magnetic moment that interacts with the orbital magnetic moment, as well as an angular momentum. These facts are not consistent with the view of an electron as a point particle, but show that it must have internal structure. Dirac's relativistic theory of the electron showed how this occurs. The electron has an angular momentum, or *spin* of one-half unit, $j = h/4\pi$, and a magnetic moment of one Bohr magneton. Therefore, the electron is twice as effective in turning angular momentum into magnetic moment. The factor 2 is called the *electron g-factor*. Actually, it is very slightly different from 2, and the tiny difference was exactly explained by the theory of quantum electrodynamics, showing that we really do know a lot of what is going on. The spin and g-factor of the electron have no classical analogues.

Closed subshells of electrons have zero angular momentum, hence zero magnetic moment. All the magnetic moment of an atom is due to the electrons in unfilled subshells. The transition metals usually have an unfilled d-shell, so they usually have magnetic moments. The orbital angular momentum of an electron that takes part in bonding, as in metals, is usually *quenched*, meaning that it vanishes, along with any magnetic moment, in the quantum-mechanical state. The other electrons may have a magnetic moment. In a crystalline solid, each atom then has a magnetic moment that can rotate freely. When an external field is applied, the moments tend to line up with it, and this is opposed by thermal agitation. The result is *paramagnetism*, seen in many substances. The net magnetic moment is proportional to the field, $M = \chi H$. The magnetic susceptibility χ is always quite small. Gases behave in a similar manner; oxygen is paramagnetic. If the atom has no magnetic moment, the application of an external field produces a small magnetization in the opposite direction, called *diamagnetism*, which is even smaller than paramagnetism.

If there are lots of magnetic moments, then when a magnetization is created in a certain direction, there is a field at any lattice point due to all the magnetic moments at other lattice points. If the crystal is cubic, or if the moments are at random points, as in a gas, then this extra field is $(4\pi/3)M$. Since it is in the direction of M, it acts to strengthen the magnetization. Should it be strong enough, it will cause all the moments to snap over into a single direction spontaneously, without the application of any external field. Alas, it is never strong enough to do this in any paramagnetic substance.

In iron metal, there is a similar interaction that is strong enough to cause all the moments to snap into a single direction. The same occurs in cobalt and nickel. This cannot be the polarizing field we have just mentioned, since it is not strong enough. What happens is that the same electrons that produce the magnetic moments are involved in the metallic binding, and the orientation of the spins of the electrons affect the binding energy of the metal. This is a much stronger interaction than the magnetic interaction of the moments, so it can bring about the desired result. The phenomenon of *magnetostriction* is evidence for this; when iron is magnetized, its shape and size may change slightly because of the coupling of the electron spins and the binding.

In iron, each atom has five d-orbitals which must

accommodate six electrons. Linear combinations of the atomic orbitals are used to generate the crystal orbitals, and then the states are filled in order of energy. Some of these orbitals are the metallic binding electrons, and some are more or less located on the ions. In iron, 0.72 electron is devoted to the delocalized metallic binding on the average.

To see how the spins affect the energy as electrons are added to the other orbitals, we can look at the helium atom. The two electrons are in the first approximation assigned to hydrogen-like orbitals in an average potential, as if they did not repel one another.

Then, the energy of the actual states is found by perturbation theory, using two-electron product states that have been properly antisymmetrized. It is a requirement that any multi-electron state must change signs if the coordinates of the two electrons are interchanged.

If the two orbital states are the same, then this can be done if the spins are opposite. If the spins are the same, the state vanishes when an attempt is made to antisymmetrize it, since the states are identical and no linear combination of them can be made to be antisymmetrical. This is called the Pauli Principle, and is a consequence of the exact identity of electrons.

In helium, the electrostatic interaction energy between the electrons is e^2/r_{12}, where r_{12} is the distance between the electrons. If the electrons are farther from each other, then the energy is lower. Grinding through the perturbation theory, the resulting energies depend on integrals like J = +" $f(1)g(2)(e^2/r_{12})f^*(1)g^*(2)\ dv_1dv_2$ and K = +" $f(1)g(2)(e^2/r_{12})f^*(2)g^*(1)dv_1dv_2$.

There is no trick here; these are just the usual "matrix elements" used in perturbation theory. J is called a *coulomb integral* and K is called an *exchange integral*. It should be clear that there are no physical exchanges either going on or implied; it is just the way it works out with antisymmetric states made of products of one-electron states. This assignment of electrons to orbitals results in two states, one of which has the electron spins opposite and energy J + K, the other of which has the electron spins the same and energy J - K. The energy is lower in the second case because the electrons are farther apart on the average than in the first case. This makes triplet states lower in helium, parallel spin states lower in iron.

The parallel spin states are lower in energy because of the effect on the electrostatic energy, not from any magnetic interactions of the spins.

This is a very common result, and holds because the electrostatic interaction is very much stronger than the magnetic. Before this was explained on the basis of quantum mechanics, Weiss postulated a fictitious magnetic field that became known as the Weiss field that would cause the magnetic moments to lock into parallelism. There is no such field, of course, and we have just explained why it looked as if one existed.

The saturation magnetization of iron at 0K is 1752 emu (22,000 gauss). In a cubic centimeter of iron, there are $6.02 \times 10^{23}(7.86/55.85)$ iron atoms. The magnetic moment per atom is found by dividing the total moment by the number of atoms, or 2.23×10^{-20} emu.

This is just a little more than two Bohr magnetons, so it appears that about two electrons are effectively unpaired at each atom. For cobalt, the number is 1.72, and for nickel 0.61. If we start with six electrons and five d-orbitals, then two electrons are paired in the lowest crystal d-orbital, two electrons are parallel in the two highest crystal d-orbitals that give the crystal binding and no net magnetic moment, leaving the two middle d-orbitals for one electron each. The fractional numbers quoted above show that there is no integral correspondence of electrons and atom orbitals, so the complete picture must be more complicated. The next atom after nickel, copper, will have no unpaired electrons and cannot be ferromagnetic. There is also no trace of ferromagnetism in the palladium and platinum groups of metals in the next two periods, so we see what a fluke it is to be ferromagnetic.

It can happen that the exchange integral is negative. In this case, the two states reverse their energy order, and the moments are antiparallel in the stable case. This often occurs in oxides, such as FeO, CoO, NiO and MnO. These are face-centered cubic lattices, which can be divided into two sublattices, such that every member of lattice A is surrounded only by members of lattice B, and vice versa. Exchange integrals can be written for A-A, B-B and A-B interactions. The A-B interactions are usually much stronger than the A-A and B-B, and this causes the minimum energy to occur when the magnetic moments on both the A atoms and the B atoms are all in the same direction, but in opposite directions on the A and B atoms. The net moment depends on the difference of the moments on A and B, which is zero, and we have *antiferromagnetism*.

If the A and B atoms have different moments, then there will still be a net moment when they are antiparallel, and the material will be spontaneously magnetized. This is called *ferrimagnetism*. Ferrimagnetic materials often crystallize in the *spinel* structure. Spinel is $MgAl_2O_4$, consisting of Al-O octahedra with Mg in the tetrahedral voids between the octahedra. Each Al is surrounded by four O's in a plane, of which it has a quarter interest in each, and one above and below, in which it has a half interest, so the unit is AlO_2^-. In the general spinel structure, the place of the Al is taken by a trivalent ion, and the place of the Mg by a divalent ion. If the trivalent ion is Fe^{+++}, we have a *ferrite*. Ferrites are spontaneously magnetized, like a ferromagnetic material, but have no metallic binding and low electrical conductivity. This means that they can be used at high frequencies, where eddy currents make the use of iron impossible. In their applications, they act just like a ferromagnetic material.

The most interesting ferrite is magnetite, Fe_3O_4. Half of the ferric iron is on the tetrahedral interstices of the spinel structure, the other half of the ferric iron and all the ferrous iron in the octahedral interstices. The moments of the ferric iron, 5 μ_B, cancel, leaving the 4 μ_B of the ferrous ions. Here we can count the electrons and states, since there are no crystal orbitals. Each unit cell contains 8 ferrous ions, so the moment per unit cell will be 8 × 4 Bohr magnetons. The experimental result is 8 × 4.07, so the agreement is excellent.

The earliest experiences with magnetism involved the lodestone, which was magnetite, named for its ability to attract bits of iron. Therefore, magnetism began with ferrites!

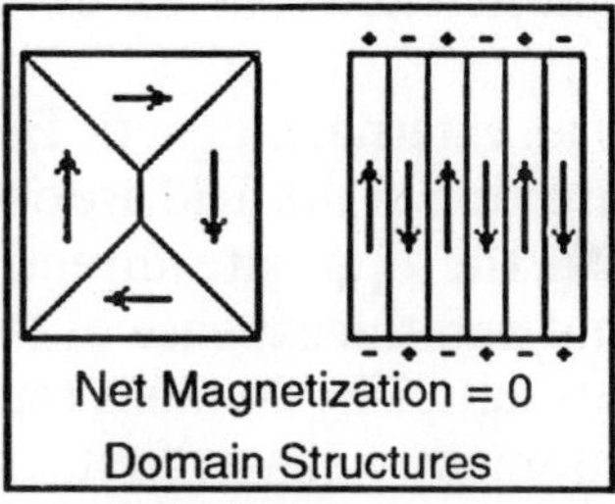

Domain Structures

Now we must explain how a spontaneously magnetized substance exhibits the magnetic properties familiar to us. If all the moments in a macroscopic sample were aligned, a very strong external magnetic field would be produced. This magnetic field represents a great deal of energy. If the moments were rearranged to reduce this external field, then the overall free energy would be reduced and the state more stable. This could be done, for example, by dividing the sample into two equal parts magnetized in opposite directions.

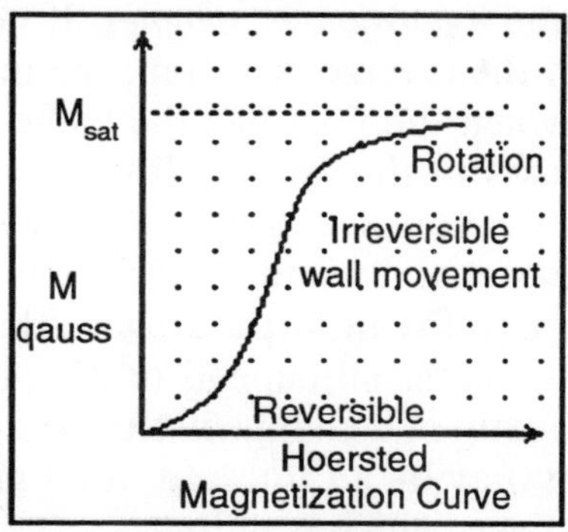

Magnetization Curve

To do this, a boundary between the two regions is required, where the moments reverse directions, and this boundary must absorb some energy. Therefore, the region will tend to divide into regions of different directions of magnetization so long as the net energy saving, field-boundaries is positive.

The regions of uniform magnetization are called *domains*, and the boundaries *domain walls* or *Bloch walls*. An individual crystallite may be a single domain or, more usually, several. The domains are arranged to reduce the external magnetic field to as low a value as possible, as shown in the diagram. Normally, no magnetic field can be detected around the sample.

When we apply an external magnetic field, those domains magnetized in the direction of the field are energy-favored over those magnetized in the opposite direction. The favored domains then expand and the expense of the unfavored, and a net magnetization results. The domains can expand by elastic movements of the domain walls at first, then as the field is increased they may jump over obstacles caused by crystal imperfections, and when the domain walls have moved as far as they can, the magnetization directions of the domains are forced into the direction of the applied field. The result is the familiar S-shaped *magnetization curve* of flux ($H + 4\pi M$ in gauss) against magnetizing field (H in oersted). If the external field is removed, the domain magnetization returns to easy directions, and the domain walls spring back as well as they can. The magnetization decreases to the *remanence*, and the sample has become a *permanent magnet*. If now an external field is applied in the reverse direction, the magnetization is reduced to zero when the field has the value of the *coercive force*, then increases in the reverse direction as it originally did in the other direction. We are tracing the *hysteresis curve* of the iron. The material does not return to the unmagnetized state except by careful manipulation. All the magnetic applications of the material depend on this behaviour. The permeability, the ratio of a change in flux to a change in H, is not constant, but may be a thousand times greater than the permeability of space, which is unity.

Permanent magnet materials are designed to offer as many pinning sites for the domain walls as possible. Once magnetized, the domain walls cannot move and the magnetization persists. Alnico V, 8 Al, 14 Ni, 24 Co, 3 Cu, with the remainder iron, is a traditional permanent magnet material. It has a saturation flux density of 12,500 gauss and a coercive force of 550 oersted. Neodymium Iron Boron and

Samarium Cobalt are modern sintered materials with excellent properties, except that they cannot stand high temperatures. Sintered magnets can easily be made in odd shapes.

"Soft" magnetic materials, where the domain walls can move easily, ideally make the magnetization a single-valued function of the magnetizing field, so that there is no hysteresis loss when the field reverses repeatedly. These materials are required for transformers and rotating machines. Crystals of pure iron have few pinning sites, so pure iron is magnetically "soft." Pure iron has a saturation flux density of 21,500 gauss and a coercive force of 0.05 oersted. Its maximum permeability is 180,000. Other alloys are designed to have as large a permeability as possible, so that only a small magnetizing field will saturate them. "Mu-metal," 18 Fe, 75 Ni, 2 Cr, 5 Cu, has a saturation flux density of 6,500 gauss, a coercive force of 0.05 oersted, and a maximum permeability of 100,000. It is used for magnetic shielding. "Supermalloy," 15.7 Fe, 79 Ni, 5 Mo, 0.3 Mn, has a saturation flux density of 8,000 gauss, a coercive force of 0.002 oersted, and a maximum permeability of 800,000.

Silicon iron as a magnetic material was developed by Sir Robert Hadfield in England, and was introduced to the United States by the Allegheny Steel Company in 1903. As silicon is added, the iron becomes a better and better magnetic material as its electrical resistivity increases, but it also becomes brittle and difficult to work. At 4.25% silicon, its resistivity is about 60 μΩ-cm, more than 7 times that of pure iron, without impairment of its magnetic qualities. The aim is to produce a steel with the lowest total core loss, eddy currents plus hysteresis. The ultimate strength of this iron is 97.9 ksi, with 5% elongation in the tensile test. At 3.25%, the elongation is still 29%, so this last percent of silicon comes at a price. The saturation flux density of the 4.25% steel is 19,700 gauss. The remanent flux density from 10,000 gauss is 8100 gauss, and the coercive force 0.4 oersted. The maximum permeability is about 7600. This represents an average working condition for the steel used in transformers. Allegheny Steel Company introduced mumetal in 1934. Silicon steel is much less expensive than the exotic alloys. To reduce eddy current losses, magnetic cores are made of thin laminations, insulated from each other.

The currents would flow normal to the laminations, so their paths are made to be of high resistance by this means. Eddy currents are, of course, just the currents that flow in conducting materials as the result of electric fields created by changing magnetic fields, according to Faraday's Law. For direct current excitation, there are no eddy currents. Transformer steel is furnished in thin sheets, from which the laminations are punched, and then assembled to form the magnetic cores.

Iron and Blood

Living cells oxidize glucose with atmospheric oxygen, and release carbon dioxide as a result. A cell bathed in water can easily dispose of the carbon dioxide, since it diffuses through the cell wall to the water in which it is very soluble, 1800 cc per litre of water. Oxygen diffuses into the cell from the water where it dissolves to the extent of 50 cc per litre. This is a small, but sufficient concentration. More oxygen would probably be deleterious. A small cell is a colloid, dominated by surface area, not volume. A cell must be small for this strategy to be successful, so most cells and simple organisms are microscopic.

When cells associate in a multicellular organism, the support of respiration by provision of oxygen and carrying-away of carbon dioxide must be arranged. Coelenterata, with an inside and an outside and specialized cells, set up a flow of water through their body that brings in oxygen, and nutrients, with the fresh water, and discharges carbon dioxide and other waste product with the efflux.

More complex organisms, such as ourselves, consist of a large assembly of cells that must respire, though out of water and packed tightly against each other. To supply oxygen and eliminate carbon dioxide, a *circulatory system* is required where a fluid transports the substances between the working cells and organs that communicate with the atmosphere, the lungs. In the atmosphere, the partial pressure of carbon dioxide is very low (0.3 torr), and the partial pressure of oxygen is high, 152 torr. If water is circulated, the soluble carbon dioxide would be transferred very well, since it would dissolve easily near the cell, and would evaporate easily when a thin layer were exposed to the air.

Oxygen, however, is a different story. Because of its low solubility, an insufficient amount would dissolve across the limited lung area to supply all the cells, who would require an absorbing area equal to their total surface area. Some vehicle is necessary that would absorb oxygen in the required amount in the lungs, and then when carried to the cells, would release it for their use. This is not a simple matter. Something that absorbed oxygen in the lungs would hold on to it at the other end as well. It is necessary to make the vehicle hungry for oxygen in the lungs, and disgusted with it in the tissues, to transport sufficient oxygen for the hosts of cells in the body.

A very intricate and wonderful mechanism has evolved to handle this job. It involves a protein, *globin*, and a *prosthetic group*, *heme*. A protein is a chain of amino acids linked by peptide groups that folds intricately in the presence of water to carry out a chemical task, using its ensemble of amino acid side chains and its structure to that end. The purpose of globin is to hold and manage heme groups that are the actual oxygen carriers.

A simple globin is *myoglobin*, in which the ingenious mechanism of the heme group is used to store oxygen in muscles, for instant release when needed. The transporter is *hemoglobin*, which consists of four myoglobin-like units carefully arranged and mechanically connected. This protein is *allosteric*, meaning that changes occurring at one point in the molecule have effects at distant points.

The purpose of the allosterism is to make binding of one oxygen at one of the four heme groups *help* the bonding of additional oxygen at the other heme groups. Oxygen bonds to hemoglobin *cooperatively*. This mechanism works in reverse when the oxygen is released. Hemoglobin with oxygen bound to its four heme groups is *oxyhemoglobin*, and without oxygen, it is *deoxyhemoglobin*. Hemoglobin not only carries oxygen, but also carbon dioxide and hydrogen ion, H^+, so it responds to the pH of its environment and the CO_2 concentration as well as to the partial pressure of oxygen.

In the lungs, the partial pressure of oxygen is high, the partial pressure of carbon dioxide is low, and the pH is relatively alkaline. All these factors encourage deoxyhemoglobin to become oxyhemoglobin. The hemoglobin is contained in erythrocytes, red blood cells, that communicate readily through their cell walls with the blood plasma. They are in constant circulation through the body, reaching the most distant cells.

Activity of a cell, especially muscle cells that are hungry for oxygen, produces carbon dioxide and acids, such as lactic acid, which make the pH acidic. The hemoglobin adds CO_2 and responds to the increased acidity and low partial pressure of oxygen by actively dumping its oxygen. The release of one oxygen makes the release of the others easier, and the oxygen pressure in the region is restored, so oxygen diffuses to all the cells in the vicinity. The whole procedure is quite automatic, and generally works very well indeed. The sensitivity of hemoglobin to CO_2 and H^+ is called the *Bohr effect*, after Christian Bohr, who discovered it in 1904.

Free hemoglobin does not have these properties; its oxygen affinity is like that of myoglobin. Within an erythrocyte, it is provided with BPG, 2,3 biphosphoglycerate, which lowers the partial pressure of oxygen at which half the hemoglobin is oxyhemoglobin from 1 torr to 26 torr. This means that a lowering of the oxygen partial pressure to 26 torr will release half the oxygen from oxyhemoglobin, so the cells will not have to gasp for air.

The effect of BPG was discovered by Reinhold and Ruth Benesch in 1967. Fetal hemoglobin, hemoglobin F, has a higher oxygen affinity than hemoglobin A. This enables the fetus to receive oxygen across the placenta. At birth, it is replaced by hemoglobin A for normal respiration through the lungs. Hemoglobin F has less affinity for BPG, and so more affinity for oxygen, than hemoglobin A.

A very wonderful thing is that all of this is now

understood in detail, even the mechanism of the Bohr effect, since the structure of proteins can be worked out by X-ray scattering. John Kendrew worked out myoglobin in 1957, and Max Perutz hemoglobin in 1959, after 23 years of effort. Myoglobin is 4.5 × 3.5 × 2.5 nm and contains about 141 amino acids and one heme group.

Hemoglobin is about 5.5 nm in diameter, and contains about 574 amino acids. The methods of protein synthesis are now known as well. An enzyme travels down a DNA double helix and makes a single-strand messenger RNA. Another enzyme starts like a zipper at one end, and travels down the mRNA, making a protein as it goes like a strip from a labeling machine. This protein then curls up, grabbing a heme group, and associates with three others to make hemoglobin. This takes place in bone marrow. There are several different kinds of chains, which make a variety of hemoglobins. The predominant adult human hemoglobin is hemoglobin A. Errors in the DNA that make defective protein chains cause recognizable diseases, like sickle-cell anemia, when the hemoglobin does not function as intended.

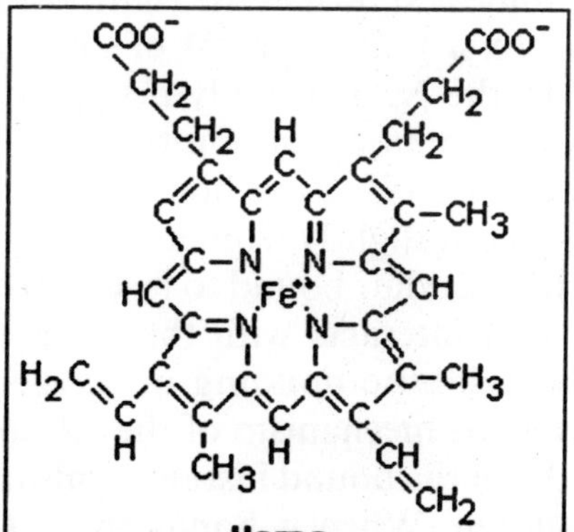

Heme

The structure of heme is shown at the right. The ferrous ion is at the centre, and is the active ingredient. Only ferrohemoglobin can bind oxygen; ferrihemoglobin cannot. The double bonds give rigidity, since there is no free rotation about them. The central part, formed of four pyrrole rings with the nitrogens holding the iron loosely, is a rigid disc. Each of the side chains can rotate about a single bond, and have rigid "hooks" that can lock them into the globin that holds the heme.

Oxyhemoglobin is transparent to light of wavelength greater than 580 nm, so it appears bright red, while deoxyhemoglobin absorbs a little in this band, appearing darker red. Both absorb strongly everything of less than 580 nanometer wavelength.

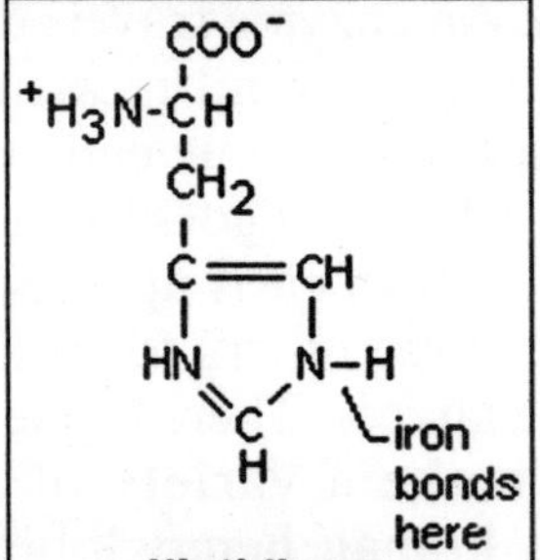

COO^-
^+H_3N-CH
CH_2
C=CH
HN N–H
C
H
iron bonds here

Histidine

When the globin winds around the heme, it places a histidine next to the iron on one side that bonds to it. The amino acid histidine is shown at the right. An H has come off the COOH carboxyl group typical of an organic acid and attached itself to the N, as usually occurs in solution. This end is part of the globin chain. On the other side there is a second histidine, but it does not bond to the iron. When the iron moves in and out slightly from its position, the motion is transmitted to the histidine bound to it, which moves a section of the protein that interacts with the other proteins of the hemoglobin to make the bonding of oxygen more or less favorable. This is the mechanism of the allosteric effect. The iron ion is a little on the bound histidine side of the heme disc in dexoyhemoglobin. When it bonds to an O_2 molecule, it moves into the plane, producing the allosteric effect. The axis of the oxygen molecule is at an angle (it binds to one of the lone-pair electrons of the oxygen). This is shown in the diagram to the right.

The iron also bonds strongly to CO, and prefers to be end-on, where all the electrons are. The body has an internal problem with CO, since it is formed by the recycling of heme at the end of the 120-day life span of an erythrocyte.

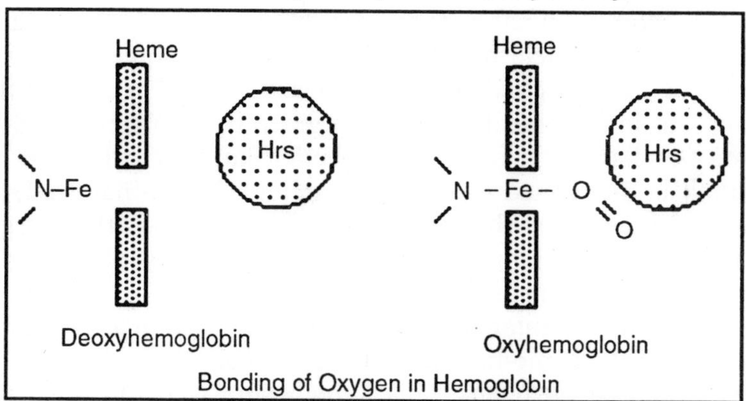

Bonding of Oxygen in Hemoglobin

There is enough CO formed to block all the hemoglobin in the blood after a while, so something must be done. The other histidine, the one not bound to the oxygen, stands just above the iron ion, and will not allow CO to bind end-on, but forces it to the side. This reduces the problem enough that only about 1% of the hemoglobin is blocked by endogenic CO. Oxygen normally binds at an angle, so the histidine does not bother it. Once CO has bonded with hemoglobin, it is very difficult to pry it off. Fortunately, few other molecules will bind to the heme.

Cysteine
Cysteine
S
Fe
Fe
S
Fe
S
S
Fe
Cysteine
Cysteine

The Fe-S Cluster in Aconitase

As heme group is also present in *cytochromes*, proteins which transfer electrons to O_2 in metabolism, converting them to H_2O as one of the end products of oxidizing food. This heme is slightly different than that in hemoglobin. *Cytochrome oxidase*, the enzyme catalyzing this reaction, also contains a heme, called heme A.

Chlorophyll has a structure very similar to heme, except that the metal held by the four nitrogens is magnesium, and one of the attached chains is rather long. This molecule absorbs in the blue and red, so is green by transmitted light, the familiar green of photosynthetic plants. It is located in small bodies called *chloroplasts*, relics of once-independent cells, where it absorbs the energy of light to supply the energy necessary for making carbohydrates out of carbon dioxide and water, supporting all the life on earth.

Another similar molecule is vitamin B12 or *cobalamin*, in which the heme-like molecule *corrin* binds a cobalt atom with the four nitrogens. This is the business site of coenzyme B12, which is necessary in purine synthesis and other duties. In coenzyme B12, one of the cobalt bonds is directly to a carbon, the only example of a carbon-metal bond in all biochemistry. Cobalamin can be synthesized only by microorganisms, typically anaerobic bacteria. It must be obtained in the diet, or from intestinal bacteria. Only about 10 μg/day is required by a human. It is so widely available that a deficiency is rare. The body takes special care to make chemicals that aid its absorption in the small intestine. If these chemicals are lacking, the result is serious *pernicious anemia*, from the resulting scarcity of coenzyme B12. The cobalt atom can be in a +1, +2 or +3 oxidation state, and all are important in the actions of vitamin B12. The cobalt in cobalamin is successively reduced to +2 and then +1, when it combines with ATP to form 5'-deoxyadenosylcobalamin, which is the coenzyme B12 essential for cell chemistry.

Iron also appears in the *iron-sulphur proteins* which are essential to aerobic life. The enzyme *aconitase* occurs in the citric acid cycle that oxidizes carbon to CO_2 in mitochondria, rearranging the citrate to isocitrate at the beginning of the cycle so it can be further processed. The prosthetic group in aconitase is the iron-sulphur group shown in the diagram at the left, composed of interpenetrating iron and sulphur tetrahedra. It is bound tetrahedrally to four cysteine residues in the enzyme. Iron-sulphur proteins also play a role in nitrogen fixation.

Iron and cobalt, therefore, are essential to life, and especially for animals with red blood. It is curious that molybdenum is also essential for the fixation of atmospheric nitrogen by bacteria, since it plays a role in a necessary protein. This protein is an enzyme converting N_2 to NH_4^+. Very little is required, but it is required. Such trace nutrients need not be supplied in any quantity greater than the need, and are usually amply available to plants and animals in their normal surroundings. The "supplements" filling the shelves in health stores and supermarkets are surplus to requirements and totally useless, the result of a combination of ignorance of nutrition with thirst for money. In fact, too much of a trace nutrient, especially metals, may be toxic. Some nonmetals, such as iodine to support thyroid function, fluorine to strengthen tooth enamel, or boron for plant growth are necessary in small amounts, but in larger amounts are also toxic. Unlike the trace metals, they may actually be deficient in the environment, and require supplementation. Iron, of course, is more than a trace nutrient, but is effectively recycled by the body and only needs to be topped up.

Chapter 9

Chromium and Manganese

Chromium (Cr, 24) and manganese (Mn, 25) are "transition metals" inhabiting the middle of the periodic table. They illustrate very well the properties of these elements, which are similar in some ways, but in others having fascinating idiosyncrasies. Nearly two-thirds of the elements of the periodic table are transition metals, in which the d and f atomic orbitals are being filled. These include the 14 "rare earths" from cerium (Ce, 58) to lutetium (Lu, 71) and the 14 "actinides" from thorium (Th, 90) to lawrencium (Lr, 103). Those at the far right: Cu, Ag, Au and Zn, Cd, Hg—are atypical, more "normal" elements in which the members of a column resemble one another. In them, the d shell is filled with its complement of 10 electrons, so d-electrons only play a role when they are promoted to higher orbitals, and do not dominate the chemistry. Chromium and manganese each have a half-filled d-shell, which is responsible for their unique properties. To the right are the triads (Fe, Co, Ni), (Ru, Rh, Pd), and (Os, Ir, Pt), all peculiar metals. Therefore, we are talking mainly about the 14 elements in the third through fifth periods to the left of these, omitting technetium (Tc, 43), which does not occur in nature. If it did, it would be like the rest. It was once hoped to be masurium, discovered along with rhenium in 1924, but masurium turned out to be an illusion.

Actually, Tc does occur in very small amounts as a fission fragment from the spontaneous fission of uranium, and some modern measurements with X-ray spectroscopy, the same method used by Tacke, have indeed detected it. However, Tacke did not *isolate* element 43 nor report any of its properties, only an amount far in excess of what could possibly be present. Whether credit for discovery should rest on such slim evidence is doubtful. Technetium was made by Perrier and Segrθ in 1937 by the $Mo^{98}(n,\gamma)Mo^{99}$ reaction (or similar) followed by beta-decay to Tc^{99}, the longest-lived isotope of technetium (2.1×10^5 y). It can now be produced in kilogram amounts by the nuclear industry.

Professor Fred Allison, a physics professor at the Alabama Polytechnic Institute (now Auburn University) announced the discovery of Alabamium (Ab, 85) and Virginium (Vi, 87), as well as the hydrogen isotope H^3, in 1931. He used the *Allison Effect*, connected with the Faraday Effect, for isotopic and chemical analysis. This proved to be an example of pathological science, and Allison was debunked by Irving Langmuitr, after he had taken in many eminent chemists. It is now estimated that in the outer mile of the continental crust there is in all only 68.6 mg of element 85, and 24.5 g of element 87 (for comparison, there are some 4000 tons of Po^{210}). Both elements have only short-lived radioactive isotopes, and Professor Allison was probably not looking in the right place anyway. Virginium had previously been claimed by Russian scientists, who proposed the names Russium and Moldavium. They are now called Astatine (At, 85) and Francium (Fr, 87), discovered in 1940 by Corson, McKenzie and Segrθ, and by Mlle. Perey in 1939, respectively. Promethium (Pm, 61) was discovered in Italy and named Florentium in 1924, then by B. Smith Hopkins at the University of Illinois in 1926 and named Illinium, but was actually discovered at Oak Ridge some years later. It does not occur in nature, since its longest-lived isotope has a half-life of 25 years.

The chemical elements show very well that they have no intrinsic qualities: they are just nuclei of differing charges with enough electrons to make them electrically neutral, and that the electrostatic forces between them are responsible for all the wonderful variety of materials. The qualities are given entirely by the structure, none by "essence." For example, it is energetically favorable (lowest free energy) for the outer electrons to rub off when the atoms are brought together and form a *metallic* bond in which they are delocalized and can more or less roam freely among the atom cores. This is responsible for the metallic lustre and the simple crystal structure: body-centered cubic, face-centered cubic or hexagonal. To this degree, all the elements look alike. We will be very interested in structure, for the understanding it can give us of the observed properties.

Both chromium and manganese are essential in the iron and steel industry, and about 90% of each is used in this application. The small resources of chromium and manganese ores in the United States are exhausted, so dependence on foreign sources for new metal is total. A small secondary (recovered from scrap) chromium production now exists, that provides about 20% of the national needs. Both metals are on the list of strategic and critical materials, with emergency stores of 4×10^6 tons of Mn and 1×10^6 tons of Cr. as of 1989. World production of chromite is a little over 10^7 tons, and world reserves are estimated at 7.6×10^9 tons. United States consumption of chromium is 0.54×10^6 tons, of which secondary chromium supplies 0.11×10^6 tons. The strategic reserves amount to about a 2-year supply, which is considerably better than the 30-day supply of crude oil on reserve. At 1989 prices, chromium was $930 per ton, manganese $215 per ton.

Chromium and manganese form many ionic complexes, often colorful ones, responsible for the colors of gems and pigments. The study of their structure is called "ligand-field theory," an interesting branch of inorganic chemistry. A few of their compounds are often used in the laboratory, like manganese dioxide, potassium permanganate and potassium dichromate. We shall take a look at the chemistry of chromium and manganese to make these things clear.

THE METALS AND THEIR PROPERTIES

Chromium was discovered and named by Nicolas-Louis Vauquelin, professor of chemistry at the Icole des Mines in Paris in 1797, whose work was confirmed by H. M. Klaproth soon after. The name is from Greek to xrwma, "the skin," by extension the "colour of the skin" and finally to "colour" in general. Xrwmata was used figuratively for "ornaments," perhaps brightly colored. Vauquelin was impressed by the vivid colors of chromium compounds, such a joy after the endless colorlessness of sodium and potassium compounds. Chromium was soon used in pigments. Lead chromate, $PbCrO_4$, found as the mineral *crocoite* is known as *chrome yellow*. Basic lead chromate, $PbCrO_4 \cdot PbO$, is *chrome red*. Chromic oxide, Cr_2O_3, is *chrome green*.

Manganese was first identified by the ingenious Swedish apothecary K. W. Scheele in 1774, but Gahn seems to have first isolated the metal later the same year. Bergmann also contributed to the knowledge of manganese. Chromium and manganese were merely chemical curiosities, without uses, except for the chromium pigments and manganese dioxide. As early as 1740, Pott had determined that pyrolusite contained no iron, and made some manganese compounds. The crystal structures of manganese metal are surprisingly complicated. There are two, α- and β-manganese. Look them up on the NRL Crystal Structures website; a link is given in the References.

The name "manganese" has an interesting history. In ancient times, two black minerals from Magnesia in Asia Minor were called Magnes, but were thought to differ in sex. The male Magnes attracted iron, and was, of course, magnetite, and gave us the term Magnet. The female Magnes did not attract iron, but was used to decolorize glass. The female Magnes was later called Magnesia, but we now know it as pyrolusite or manganese dioxide. In the 16th century, it was called Manganesum by glassmakers, while the alchemist Camillus Leonardus called it Alabandicus (from a location where it was found), and it was also known as brownstone (Braunstein). Mercati called it Manganesa, and finally the metal isolated from it became known as Manganese (German: Mangan). The name Magnesia was then free to be applied to white magnesium oxide, which provided the name Magnesium for the metal. Magnet, manganese, magnesia and magnesium all have the same etymological source. The most remarkable physical properties of chromium are its hardness and its passivation by a thin transparent surface layer of oxide.

Ordinary chromium metal has a hardness of Mohs 9, equivalent to corundum, inferior only to diamond. It may be the hardest metal, absolutely unscratchable by any other. It cannot be scratched by a file, which has a hardness of about 6.5.

Very pure chromium, and the porous chromium of decorative chrome plating, may not be quite as hard, but they are still up there. This means also that chromium will take a superb polish, and the passivation means that this surface will not corrode.

A few other metals, such as aluminium and nickel, can also be passivated, and non-metals like silicon as well, but they are not nearly as hard. The "stainless" in stainless steel results from this same property, when surface chromium atoms form a passivated surface that resists chemical attack.

Manganese, on the other hand, has a hardness of 5, about the same as a knife blade, so it is hard, but not exceptionally so. Most transition metals are characteristically this hard. Like iron, it rusts, and perhaps even more readily. Along with its hardness it is brittle, a property shared with chromium, which means that neither pure metal is suitable as a material of construction. Both metals are silvery on a fresh surface, which we noted above was a common characteristic of metals with free conduction electrons. The electrical resistivities of chromium and manganese are given in Lange as 2.6 and 5 $\mu\Omega$-cm; another source gives 13.1 and 5, and the Sargent-Welch periodic table 12.9 and a rather unbelievable 144. For comparison, copper's is 1.68.

The melting points of chromium and manganese are 1857°C and 1244°C, and the boiling points are 2642°C and 2040°C. High melting and boiling points are typical of transition metals, tungsten having the highest of all, with tantalum and rhenium not far behind. The densities are 7.19 and 7.43 g/cc, about the same as those of iron and other familiar metals in the third period.

The electron configuration of chromium is $Ar3d^5 4s$. There are six electrons outside an argon kernel, and the energies of the 3d and 4s orbitals are not very different. Chromium is found with oxidation numbers of 0, +2, +3 and +6. This certainly does not mean that bare ions of the corresponding charge exist, but that the charged chromium kernel produces an attractive field for negative ions, such as O^{--}, or for the negative ends of polar neutral groups, such as H_2O.

Structures may be different in aqueous solution, indeed in acid or basic solutions, and in crystals. Bonds may be ionic or covalent, or of any intermediate character. Whatever gives the minimum free energy is what happens. Oxidation number zero is, of course, metallic chromium. Cr^{++} is called *chromous*, usually quickly oxidized to Cr^{+++}, or *chromic*. In acid solution, these ions coordinate 6 water molecules around them in an octahedral arrangement, attracting the lone pairs.

The chromous ion is blue, the chromic ion violet. In basic solution, four hydroxyl ions are bound tetrahedrally, to give $Cr(OH)_4^{--}$ [$Cr(OH)_2$ + $2OH^-$], or $Cr(OH)_4^-$ [CrO_2^-, chromite ion, which is green]. The +6 state coordinates four O^{--} ions to form CrO_4^{--} [chromate] in basic solution, or Cr_2O7^{--} [dichromate] in acidic solution. The dichromate ion is two tetrahedral chromate ions joined at one vertex. There is no change of oxidation number between chromate and dichromate, but the colour changes from yellow to orange. The situation is a little messier than this in detail, since H_2O and OH^- ions can be mixed, but this gives the general appearance and behaviour of the soluble compounds. Zinc metal can be used as a reducing agent, and hydrogen peroxide as either an oxidizing or a reducing agent, in working with chromium ions. Soluble chromium compounds are poisonous.

When dichromate and concentrated sulphuric acid are mixed, the reaction gives a sulphate and chromic acid, which is dehydrated to red chromium trioxide, a powerful oxidizing agent. This is the familiar *cleaning solution* for laboratory use. To make it, mix 800 cc of technical concentrated sulphuric acid with a solution of 92 g of technical $Na_2Cr_2O_7{\cdot}2H_2O$ in 458 cc of tap water (or corresponding proportions). Leave the precipitate alone; it is part of the solution. It should be used warm or hot, and only after preliminary cleaning with detergents has left only the slightest residue. It is expensive and hazardous, and not recommended for home use. When the solution turns greenish, it is exhausted and should be discarded, followed down the drain by ample water to dilute it, since it is strongly acid, but not hazardous in a dilute condition. It will help keep the drains clear.

Manganese has one more electron, and shows even more oxidation states: 0, +2, +3, +4, +6 and +7. The +2 manganous ion is rather stable and not easily oxidized in neutral or acid solutions. The manganous ion is colorless in aqueous solution, but some manganous salts are pink. In a basic solution, white $Mn(OH)_2$ is formed, but it rapidly oxidizes to a +3 oxide. The +3 manganic ion quickly disproportionates, some reducing to +2 while an equal amount oxidizes to +4. The +3 state can be preserved only by complexing with CN^- or $C_2O_4^{--}$ or a similar group, or by forming an insoluble salt. The oxalate is a *chelating* ligand that binds at two points, a favored structure. *Ligand* is just a name for any group of atoms that is bound to the central ion. The principal compound of the +4 state is MnO_2. It is an oxidizing agent when the manganese is reduced to Mn^{++++}, and a reducing agent when the manganese is oxidized to +6. Heating the dioxide with a basic substance in air makes dark green manganate ion, MnO_4^{--}, with +6 manganese. When acid is added, the manganate disproportionates to MnO_2 and +7 manganese in the deep violet permanganate ion, MnO_4^-. Thus, K_2MnO_4 is potassium manganate, while $KMnO_4$ is potassium permanganate, the familiar soluble violet salt. Permanganate is the strongest common oxidizing agent, because it is readily reduced. The disappearance of the violet colour gives the end point in titrations.

Manganese dioxide is used as a depolarizer in dry cells. By "polarization" is meant the appearance of a layer of nonconducting adsorbed molecular hydrogen. Manganese dioxide removes this by oxidizing the hydrogen to water, perhaps by the reaction $MnO_2 + H_2 \rightarrow MnO + H_2O$. It can also be used to oxidize the chlorine in HCl to Cl_2 in the laboratory production of chlorine. It is also a catalyst in the thermal decomposition of potassium chlorate, $2KClO_3 \rightarrow 2KCl + 3O_2$, the usual laboratory source of oxygen. I have not heard of a theory of why this takes place, or if the manganese dioxide must be prepared a certain way so that it has a large surface area.

Trivalent transition metal ions can replace trivalent aluminum ions in corundum, Al_2O_3. Electron transitions in these ions, unlike in the colorless aluminum which does not have such electrons available, can absorb strongly in the visible spectrum and give bright colors. Cr^{+++} absorbs the blue and green, leaving only a deep-red colour, making a ruby. Ti^{+++}, on the other hand, absorbs the green and red, making the blue sapphire. In beryl, $Be_3Al_2(SiO_3)_6$, Cr^{+++} absorbs blue and red, creating the green emerald. The crystal environment of the ion is different in beryl than in corundum. Many other clear crystals are colored by transition-metal ions. The amounts required are so small that in many cases the responsible ions have not been identified.

Ruby absorbs light over a broad spectrum. In an ordinary crystal, the energy is mostly dissipated by phonon interactions, and mainly the result of the absorption is seen. When excited by ultraviolet, however, the red emission can often be seen, especially in artificial ruby intended for lasers. In a laser, the ends of the crystal may be carefully polished parallel to make a resonant cavity for red light (or external mirrors can be used). When the ruby is "pumped" with white light, such as that from a xenon flash lamp, it may emit stimulated coherent radiation in the range 693-705 nm, a very deep red. This radiation comes from the Cr^{+++} ions.

Mineralogy and Metallurgy of Chromium and Manganese

Chromium and manganese are both widely distributed in the earth's crust. The abundance of manganese is 0.096%, and that of chromium is 0.037%. Chromium has not been detected in sea water, but manganese is present to about 1-10 ppb. Their compounds are very insoluble, so this scarcity is not surprising. Chromium occurs almost entirely as ferrous chromite, $Fe(CrO_2)_2$, which is called *chromite*, a sample of which is shown in the photograph (see References for the source of this copyrighted image).

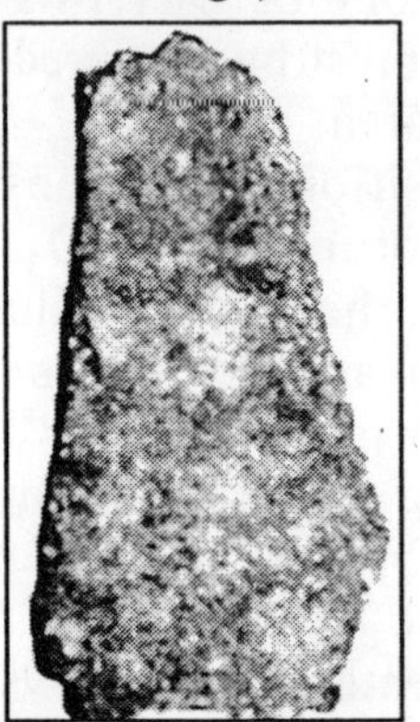

This substance crystallizes early out of a basic magma, and so is found associtated with peridotites, serpentine, olivine and other ultrabasis rocks. It is heavy and can be concentrated by gravity methods. The largest producers of chromite are Kazakhstan and South Africa, responsible for about 75%. India, Zimbabwe, Turkey and New Caledonia (more on New Caledonia below) are other sources. It is a hard, black crystalline substance that leaves a brown streak on unglazed porcelain.

Some chromite is ground, mixed with a little lime and a binder, formed into bricks and baked to form a refractory furnace lining. Chromite does not react with either silica or magnesite refractories, and so is called a *neutral* refractory, of which it is practically the only example, and, so is indispensable.

Chromite was first mined in the United States in 1828, in Pennsylvania and Maryland. In World War II, there was production from California and Oregon. Chromite is not currently mined in the United States, but about 20% of current consumption is met by *secondary* chromium, that is, chromium recovered from scrap stainless steel. 18-8 stainless steel can be directly charged into electric furnaces that make most stainless steel. The nickel in scrap steel is more valuable than the chromium. To form chromates, the chromite is first roasted with sodium carbonate in the presence of air.

The result is ferric oxide, carbon dioxide and sodium chromate. The chromium is oxidized from +3 to +6 in this reaction. When the chromate is dissolved in dilute sulphuric acid, sodium sulphate and sodium dichromate result. The sodium sulphate is less soluble than the dichromate, so it is crystallized out, and the dichromate recovered.

When sodium dichromate is heated with sulphur, sodium sulphate and green chromic oxide results, which is insoluble and remains when the sulphate is washed out. Chromic oxide and aluminium powder are mixed and ignited (Goldschmidt or Thermite process) to produce molten chromium metal of 99.5% purity. If dry sodium chromate is treated with concentrated sulphuric acid, sodium sulphate and chromic acid, H_2CrO_4 are produced. The chromic acid is immediately dehydrated by the sulphuric acid, and red CrO_3 precipitates. Chromium trioxide is used in chromium electroplating, where it is known as "chromic acid."

To make ferrochrome, the alloy used in making alloy steels, chromite can be smelted with carbon in a blast furnace. Ferrochrome should contain about 60% Cr and as little carbon as possible. It is possible to decarbonize ferrochrome in a reverberatory furnace under a basic slag of lime and feldspar. The lower the carbon, the higher the value of the ferrochrome. Although the U.S. once produced most of its ferrochrome, it is now mostly imported. The price of chromite is now about \$75 per metric ton, ferrochrome \$1000 per metric ton, and metallic chromium \$7500 per metric ton, about \$3.30 per pound. Nickel, by comparison, is 5 to 8 times as expensive.

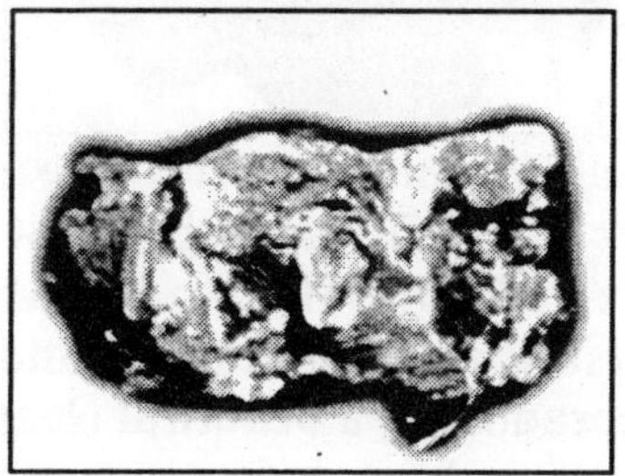

Manganese, by contrast to chromium, occurs widely disseminated in igneous rock, and so in soils and sedimentary rocks. The only economic ores are those that have been concentrated by *supergene* enrichment, dissolved by descending waters and deposited in favorable locations. The most common mineral is *pyrolusite*, which is mainly MnO_2. In its crystalline form, which is very rare, it is hard and black.

A crystalline specimen is shown at the left. (See References for the source of this copyrighted image). Usually pyrolusite is found in soft masses of colloidal particles, hardness 1-2, and staining the hands like charcoal.

An impure form containing mixed oxides is called *wad*. The name pyrolusite comes from the Greek for "fire" and "dissolve" since if added to molten glass, it oxidizes iron from ferrous to ferric, "washing out" the green colour and making clear glass. It often contains iron impurities, and so may be slightly magnetic, which led to an early confusion with magnetite, and the "magnetic" name of manganese ("magnanese" was awkward). The shales of South Dakota contain some manganese, about 1%, and means of beneficiating this to 18% by leaching have been devised. The product can be used to extend more concentrated ore. An ore of about 40% metallic manganese is required for present processes.

Manganese is also found in several rare minerals, such as pink *rhodochrosite*, $MnCO_3$, and rhodonite $MnSiO_3$, and black manganite, MnO(OH), and alabandite, MnS. Rhodochrosite crystals resemble calcite, and those found in Lake County, Colorado, are a beautiful clear pink. They are the "state mineral" and are sometimes cut as gems. An attractive specimen of rhodochrosite is shown at the right (© Amethyst Galleries). These minerals can sometimes be ores of manganese where they occur as a by-product, as rhodochrosite was at Butte, Montana, but are seldom found in sufficient amounts to be worth mining.

In the semi-arid country of southeastern Utah, red sandstone cliffs are often seen to be covered by a black film. In some places, *petroglyphs* have been created by the early inhabitants (and some by recent more barbarous characters), scratched through the black surface. This film is *desert varnish*, and is made black by its content of manganese dioxide. Infrequent moisture is first absorbed by the rock, then is drawn to the surface as the moisture evaporates, leaving dissolved substances behind, among which is the manganese, as well as iron from the cement of the sandstone. Manganese and iron oxides form a tenacious layer on the surface of the rock, which is the desert varnish. As the rock weathers, new varnish appears to recruit the old. This shows again how common manganese is in the environment, though in very small concentration.

Manganese nodules, or *ferromanganese concretions*, in the deep ocean can contain up to 36% Mn, but 30% is usual in rich nodules. These nodules are found in both Atlantic and Pacific, but principally in the Pacific. One locality is the Manihiki Plateau east of Samoa, from 5°-15°N, 120°-160°W, at depths of 12,000-18,000 feet. The nodules are concretions of 1" to 9" diameter, lying in diatomaceous ooze. The concretions begin on a shark's tooth or whale's earbone, which are very insoluble, or other nucleus, and grow about 1 mm in a million years. They grow from their bottom surfaces, and are kept moving by benthic organisms. There can be 100,000 tons of nodules per square mile. They also contain copper, nickel, cobalt and tungsten. Mining, apparently mainly for copper and nickel, was to begin in 1977-1980, but the four American companies then involved have abandoned their plans. There are complex legal and environmental obstacles to be overcome in nodule mining in the deep sea. The nodules were discovered on the pioneering oceanographic cruise of

HMS Challenger in 1873-1876. Russia and South Africa produce about 85% of the world's pyrolusite. Large new sources have been found at Moanda, Gabon; in the Kalahari Desert of north Cape Province, South Africa; and at Groote Eylandt, Northern Territories, Australia. Brazil, New Caledonia and even Fiji also have manganese ores. In the United States, some pyrolusite was found in South Carolina, but the reserves there have now dwindled to nothing. World War II caused the small domestic reserves of richer manganese ores to be intensively exploited, so that they no longer are an economic source. The United States currently mines no manganese.

Pyrolusite cannot be beneficiated to any great degree because of its powdery nature, so is used as it comes. Pyrolusite can be reduced with carbon in a blast furnace. Since the ore also contains a large amount of iron, the result is a white cast iron called *spiegeleisen*, or "mirror-iron" that can apparently be polished to make a mirror. This alloy contains 15-20% manganese and 5% carbon, and is inexpensive because it uses available raw materials and does not involve a chemical separation. This was the alloy first used by Mushet in 1856 to make the Bessemer process successful. A higher-quality product is *ferromanganese*, containing 75 to 81% manganese and 5% carbon. The pyrolusite must be purer, and a more limestone is used as a flux. Nevertheless, the manganese loss in the slag and vapour is as much as 15%. Ferromanganese can be decarbonized by melting with manganese dioxide in an electric furnace; the MnO_2 burns off the carbon as CO_2, and the carbon can be reduced to less than 1%. Carbon-free manganese metal can be produced from pure MnO_2 by the Goldschmidt process for use in softer high-manganese steel.

Chromium and manganese are both obtained from single ores, chromite and pyrolusite, that are widely distributed enough that monopolies cannot easily be formed. The United States, Europe and Japan have now exhausted the small resources they once had of both chromite and pyrolusite. The world supply is not in immediate danger, but the rate of exploitation of inexpensive chromium and manganese is high, and will not last a millennium, a very short time in geological terms. The U.S. strategic reserve of chromium has been sold off after the dissolution of the Soviet Union in 1991.

Electron Phases in Manganites

Manganite, as we have mentioned, is the mineral containing MnO(OH). The same name has been given by solid-state physicists to a series of crystalline compounds like $La_{1-x}Ca_xMnO_3$, where 0 d" × d" 1. The lanthanum can be replaced by any trivalent rare earth, and the calcium by any alkali metal. For × = 0, we have $LaMnO_3$, in which the manganese is +3, and for × = 1, we have $CaMnO_3$, in which the manganese is +4. The electronic configuration of +4 Mn is $Ar3d^5$, with exactly a half-filled d shell, while with +3 Mn there is an extra sixth electron. By controlling the relative amounts of La and Ca, we can control the density of the sixth electron, and so the electronic properties of the material. The manganese will be a mixture of +3 and +4 ions, which can arrange themselves randomly or in an ordered structure.

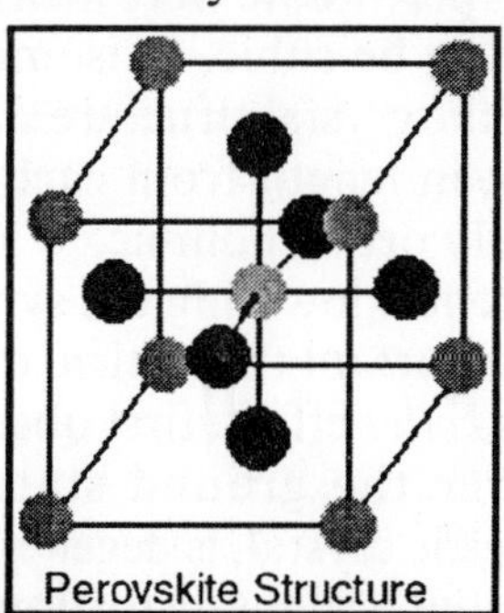

Perovskite Structure

The manganites have a *perovskite* structure, of which a unit cell is shown at the right. Perovskite is the rare mineral composed of $CaTiO_3$. It is hard (5.5), heavy (4.0 g/cc), varying from black with a metallic lustre to yellow. It was named after the Russian mineralogist L. A. Perovski (1792-1856) when first described in 1839. It has been found at Magnet Cove, Arkansas, among other places. The violet circle represents the Ti in perovskite, or the Mn in manganites. It is surrounded octahedrally by six oxygen ions, represented by blue circles.

The charge is neutralized by eight cations at the corners of a cube. Since the unit cell has a half-interest in each oxygen represented, and an eighth-interest in each cation, while the manganese is fully owned, the formula is $CaMnO_3$, if the cations are Ca^{++} and the Mn is +4.

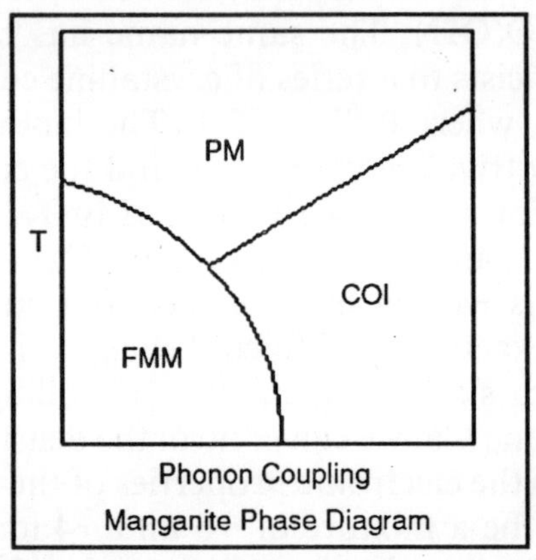

Manganite Phase Diagram

When crystals of perovskite were measured optically, the symmetry was found to be cubic, or isometric. However, X-ray measurements of the crystal structure showed that the axes were slighly different in length from each other, so that the symmetry was actually orthorhombic.

This small difference, lowering the symmetry from cubic to rhombic, has an important theoretical explanation. It is an example of the *Jahn-Teller* effect, first observed in symmetric molecules. Whenever the ground state of a symmetric molecule, or a symmetric crystal, is degenerate, a lower energy can be obtained through a slight distortion that decreases the symmetry.

The decrease in energy is E_{JT}, the Jahn-Teller stabilization energy. The sixth electron in a manganite, if localized on a particular ion, can stabilize the crystal by the Jahn-Teller effect. This is often called an electron-phonon interaction, though the phonons appear to be zero-frequency.

Another thing that the sixth electron could do is to become *delocalized* and wander about the crystal freely. This also stabilizes the crystal, since the kinetic energy of an electron free to wander is lower than if it is localized in a small volume. This can be understood by the uncertainty principle, ΔpΔx d" h/4π. If an electron is confined, then Δx is small, so Δp must be large, and vice-versa. The 1s electrons close to the nucleus of, say, a manganese ion with nuclear charge +25, are confined to a very small volume, but are also necessarily moving at a very high speed. This effect is partly responsible for the popularity of metallic lattices.

A third factor in the crystal structure of manganites is electronic magnetism. The three electrons in +4 Mn occupy the three degenerate orbits belonging to a triply degenerate T representation of the cubic group made from the d orbitals. The remaining two orbitals belong to a doubly-degenerate E representation and are at a higher energy (higher by the ligand-field splitting). The spins are parallel, since this allows the electrons to be as far apart as possible. Purely magnetic interactions tend to align spins parallel, but this is not a large effect, and most of the stabilization energy is an "exchange" effect arising from electrostatic repulsion. The three electrons form a *core moment* on each Mn ion.

All of these factors compete with each other, and with the disordering effect of thermal agitation, to determine the electronic state of a manganite, and the battle is indecisive, since all are of comparable strength. The effect of thermal agitation appears in the term -TS in the free energy. At higher temperatures T, states of greater entropy are favored over more ordered ones. Three principal kinds of phases can be recognized, and a schematic phase diagram is shown at the left. The abscissa is the relative strength of the Jahn-Teller stabilization with respect to delocalization. If the Jahn-Teller stabilization is small, then the electrons form a degenerate Fermi gas the usual way, and the crystal is a conductor. The exchange interactions between the core moments, mediated by the free electrons, align the core moments ferromagnetically, so the crystal is *ferromagnetic*. This phase is denoted FMM. If the Jahn-Teller stabilization predominates, then the sixth electron is localized, and the crystal becomes an insulator. The exchange interaction is such that a crystal with an intermediate value of x, say × = 0.5, becomes *antiferromagnetic*, with the core spins alternating in direction. This phase is called COI, for "charge-ordered insulator," and the electrons may well be described as an "electron solid."

As the temperature is raised, a disordered system where some electrons are localized and others are not, and the core spins are not locked parallel or antiparallel, is formed. This structure is *paramagnetic*, and is represented on the phase diagram by PM. The phase transitions, since they represent abrupt changes in entropy, are the usual first-order phase changes with latent heats. Since the system is quite complex, there may also be additional phases of the same general character, showing second-order phase transitions. In the PM structure, resisitivity is high, and the electron state may be described as a "liquid."

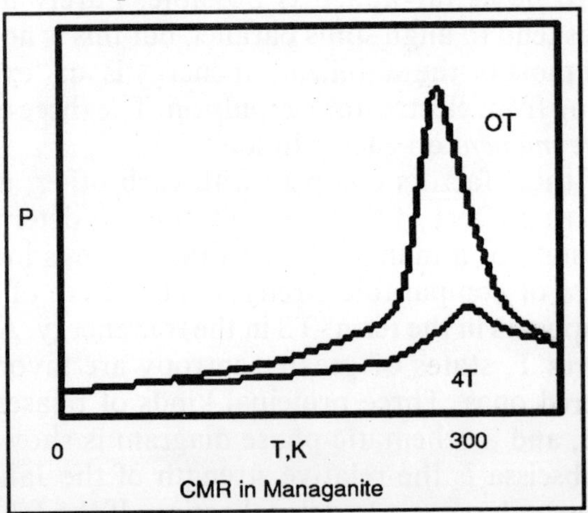

CMR in Managanite

In manganites, the indecisive battle of the states may result in the simultaneous presence of different phases in the same sample. This can occur in polycrystalline samples, or where there is incomplete thermal equilibrium. A sample of $Bi_{0.24}Ca_{0.76}MnO_3$ investigated by scanning tunneling microscopy (STM) showed a phase boundary between FMM and COI, or at least between surface phases of these types, the first disordered and conducting, the second ordered and insulating. The V-I characteristics of the individual regions were also measured, and had the expected forms. In the ordered state, the +4 and +3 Mn alternated in a unit cell twice as large as the basic one. Another imaging of a 5μm square region by magnetic force microscopy (MFM) clearly showed patches of FMM and COI. Such "mesoscopic texture" is exiting to those who study phase changes.

Another curious effect arises in the magnetoresistance of the crystals. Transverse magnetoresistance is generally seen as an increase in resistivity in the presence of a perpendicular magnetic field. For a single type of carrier and constant relaxation time, the Hall field can cancel any effect of the magnetic field on resistance with a drift velocity, but this is not possible in more complex cases, and the resistance increases with the magnetic field. The effect is small, but can be used to study the Fermi surface.

In the case of manganites, a magnetic effect called Colossal Magnetoresistance, or CMR, appears. It is called "colossal" because in some cases the difference can be a factor of 10^7, since one phase is insulating. In the transition from FMM to PM, the resistivity generally spikes as the electrons become localized, decreasing at higher temperatures as the "electron liquid" beomes less viscous. This behaviour is shown in the diagram by the curve marked "0 T." A very strong magnetic field will keep the core spins aligned, which facilitates the movement of the electrons, and so the resistivity does not rise as much, as shown by the curved marked "4 T," for a field of 4 tesla. Producing a field this strong takes special measures; it cannot be done with mere iron.

Chromium and Manganese in Steel

The Bessemer steel-making process burns the carbon out of molten pig iron by blowing air through it. In a few minutes and with a shower of sparks, tons of low-carbon steel are produced. The alternative was hours of tedious "puddling" to make wrought iron. Bessemer steel brought the age of low-cost steel to engineering in 1855. It was difficult to judge the end-point accurately, so excess oxgen was dissolved in the steel, which had a very deleterious effect. The Bessemer steel was hard and unworkable, causing Henry Bessemer to despair. Robert Mushet solved the problem in 1856 by throwing a little spiegeleisen in the blown iron. The manganese removed the oxygen by forming harmless MnO_2, and the carbon in the spiegeleisen brought it up to the desired concentration of something less than 1%. Manganese also reduced the deleterious effects of sulphur and phosphorus by preventing them from concentrating at the grain boundaries. Sulphur makes steel crumble when it is rolled red-hot, whild phosphorus makes it brittle when cold. The addition of less than 2% manganese made Bessemer steel widely acceptable. No matter how steel is now made, a little manganese is an essential addition for deoxidation and the other beneficial effects. About 16 pounds of manganese metal are required for each 2000 pounds of steel. In the 1980's, new methods of making pig iron reduced the manganese requirement by about 20%.

Steel is not considered an "alloy" steel unless it contains 2% or more of manganese, more than 0.1% molybdenum or vanadium, 0.3% tungsten or cobalt, or 0.5% chromium or nickel. This statement gives an idea of the metals that are usually added to iron to give desirable properties, such as strength, toughness, ease of heat treatment, corrosion resistance, machinability and so forth. They are all transition metals of the kind we are discussing, or else members of the Fe-Co-Ni triad. Carbon, up to a few percent at most, is essential for giving steel its desirable properties, through the formation of iron carbide, Fe_3C, cementite, hard and white, which affects the microstructure of the metal. Low-carbon steel, with less than 0.83% carbon, is a mixture of ferrite or α-iron, iron with a little dissolved carbon, and pearlite, the eutectoid mixture of ferrite and cementite. Ferrite is magnetic and hard, with a body-centered cubic crystal structure. When steel is heated to red heat, about 910°C for pure iron, the ferrite changes to soft, face-centered cubic austenite or γ-iron, which can easily be worked, and is non-magnetic. Iron ceases to be magnetic above 769°C, the Curie Point. The heat treatment that is possible because of these phase changes is one of the principal advantages of steel.

The addition of chromium to steel hardens it and impedes the transition to austenite, raising the temperature of the phase change. Chromium and ferrite are both body-centered cubic, while austenite is face-centered. Nickel (and manganese) encourage austenite, and lower the temperature of the phase change. In fact, enough nickel can be added to make austenite stable at room temperature, and the steel remains nonmagnetic. Nickel and chromium largely remain in the ferrite, but chromium forms a little carbide, which explains its hardening effect. The effect on austenite can be minimized by adding both chromium and nickel together, since this bad effect tends to cancel, while the good effects of the additions tend to add. Chromium is added in the form of ferrochromium and nickel as ferronickel.

The first alloy steel was Sir Robert Hadfield's high-manganese steel, created in 1882. This austenitic steel contains 12.5% manganese and 1.2% carbon, and is used quenched from 1050°C so the carbides remain in solution. It is hard and has a tensile strength of 123,000 psi. It work-hardens to resist abrasion to a remarkable degree, so that it is used in crushers and dredgers, and in railway track fittings that must resist great wear. Alloys with less magnesium, around 1.5%, and lower carbon, are inexpensive substitutes for nickel-chromium steels.

More than about 10% chromium renders iron or steel corrosion-proof, from the passivation of the chromium at the surface. A "stainless iron" with 13% chromium and less than 0.05% carbon is used for sinks and food containers. A higher-carbon alloy of 0.3% carbon is a stainless steel, and is used in stainless-steel knives. This alloy is martensitic with small particles of carbide. Most stainless alloys with chromium only are ferritic in structure and magnetic. The most commonly used stainless steel is Type 302 or "18-8 stainless steel." Its analysis is 17%-19% Cr, 8%-10% Ni, with C > 0.08%, Mn d"2% and Si d" 1%. It is austenitic and nonmagnetic. It is subject to work-hardening, so is not suitable for deep-drawing or machining, but is easily worked cold and welded. Extra-low carbon stainless steels (Types 304, 304L) avoid precipitation of carbides near welds. The strongly-worked metal may become magnetic.

An alloy with 12.5% each Cr and Ni is suitable for deep-drawing. An 18-8 steel with 1.5% Mn and a 0.25% S (Type 303) is free-cutting. Type 302 has excellent oxidation resistance up to 870°C, and resists corrosion by all common chemicals except hydrochloric acid and sauerkraut brine. In particular, it is not attacked by nitric or sulphuric acids. It gives a deep green solution when dissolved in hydrochloric acid. Type 301 is a 17-7 stainless, cheaper because less nickel and chromium are used, but is not as corrosion-resistant as Type 302.

Small amounts of manganese added to aluminium harden it without affecting the corrosion-resisting properties. Duralumin contains 4% Cu, and 0.5-1.0% each of manganese and magnesium. Another alloy has 1.0-1.5% manganese and 0.2% copper. Brass with up to 3.5% manganese becomes "manganese bronze" with superior strength characteristics, equivalent to mild steel, while the brass can still be cast the usual way. Manganese is very commonly used to harden magnesium, in nearly all of the Dowmetal alloys. Dowmetal F is 4% Al, 0.3% Mn and 95.7% Mg. Nichrome IV or Chromel A is 80% Ni, 20% Cr (resistivity 103 μΩ-cm). There are many alloys called "nichrome" of varying constitutions. One is Ni 60%, Fe 24%, Cr 16%, C 0.1%. Resistance alloys may be designed for small temperature coefficient of resistance, for stability at high temperatures, or for maximum resistivity.

The importance of hard steels for making cutting tools for lathes and milling machines, for punches, shears and dies cannot be overestimated. The steel must not temper (soften) when hot, must resist oxidation, and should be easy to manufacture. Fast cutting means hot tools, and was the origin of the term "high speed steel." Robert Mushet was the first to make a tool steel when he made a manganese steel in 1861 with the accidental addition of 6% tungsten. The manganese was replaced by chromium around 1881, and more tungsten was added. In 1900, the first modern high-speed steel with 18%

W and 4% Cr was exhibited cutting at red heat. Vanadium was introduced in 1906, and then molybdenum as a substitute for tungsten. W, Mo, Nb, and V all form very hard carbides, which are embedded in a tough matrix. V_4C_3 is a particularly abrasion-resistant carbide. High speed steel is 75% Fe, 18% W, 6% Cr, 0.3% V, 0.7% C, and can be forged and heat treated. The matrix is Fe-Cr. Stellite 3 is 55% Co, 20%-23% Cr, 15-20 W, 3%-5% Fe, 1.5%-4% C. Stellite must be cast, and ground to shape. Molybdenum steel has been specially popular in the United States, because of a lack of domestic sources of the other alloying metals. The famous mine at Climax, Colorado has now been closed, however. The Vickers Pyramid Hardness Number (VPN) of high-speed steels runs from 800-1000.

NEW CALEDONIA

Nouvelle Calıdonie is a French overseas territory in the southwestern Pacific, a slender island 250 miles long and 30 miles wide oriented NW-SE, about 1000 miles east of Australia and the same distance northwest of New Zealand, at latitude 21°S and longitude 164°E, just north of the Tropic of Capricorn. The Loyalty Islands are scattered along the east coast, with Fiji another thousand miles to the northeast, on the way to Samoa. New Hebrides, now Vanuatu, lies to the north. The main island, La Grande Terre, is one long mountain range, La Chaξne Central, from Mt. Panıı, 5413', in the north to Mt. Humboldt, 5360', in the south. There are coastal plains in the west, but in the east the mountains come down to the sea.

The three main Loyalty Islands are former coral atolls that were elevated in the Tertiary and are now flat round plains ringed by cliffs. One island, Ouvıa, the northernmost, largely retains the atoll configuration. Lifou and Marı are the other two. A number of smaller islands are also in the territory. Of these, Nou and Pines were penal colonies from 1864 to 1897. The mixed population of La Grande Terre is less than 200,000, of which about 50,000 inhabit the capital, Noumıa, at its southern tip. An aerial view of Noumıa is shown in the heading (source: www.sponline.com; see References).

The islands were inhabited by Melanesians at least as early as 2500 BCE. Polynesians arrived in the 12th to 17th centuries, apparently from the northeast, and settled the Loyalties. The island was discovered and named by Capt. James Cook in 1774. Napolıon III annexed the territory, a land of cannibals, in 1853. Nickel was discovered in 1863.

Prisoners from the Commune de Paris were sent there in 1872. Mining began in the late 19th century, and southeast Asians came to work the mines, mainly as indentured servants. It was on the sidelines in World War II, but a large U.S. base was established in 1942. The population is 43% Melanesian, called Kanaks, and 37% of European origin.

The remainder are descendants of the indentured laborers. Indentured servitude was abolished in 1951, and many indentured miners returned to their native lands. Nouvelle Calıdonie became an overseas territory in 1946. The various ethnic groups have not integrated, and struggle for supremacy. The Kanaks want independence, the Europeans to remain associated with France. A referendum in 1987 (boycotted by Kanaks) decided for remaining in the French commonwealth, although the surrounding island groups, such as Vanuatu (New Hebrides) and Tonga (Friendly Islands), are now largely independent nations. There is still agitation for independence, of course, and rumblings continue. The direct dialing code is 687, and its time zone is +11, one hour later than eastern Australia. Most of the rugged 6530 square miles are unsuitable for agriculture, but some of the typical agricultural crops of the region—coconuts and yams—are produced, chiefly along the west coast.

The island lies southwest of the New Hebrides trench, in a region where the Indian plate is subducting beneath the Pacific plate, scraping off the mountains of New Caledonia and creating many volcanic islands and atolls. The igneous rocks have provided a wealth of minerals: Ni, Cr, Fe, Co, Mn, Ag, Au, Pb and Cu are all mined. The island is the principal source of nickel, which is arsenic-free, after Canada. The lateritic nickel mines were opened in 1875, and there was a Nickel Boom, 1968-71, after Canadian labour problems caused a nickel shortage. Chromium and manganese are also produced here, though not often shown as a source in references. I do not know the status of the mines and the reserves, and the website scarcely mentions mining. Petroleum, coal and coke are imported to fuel the mineral industries. I was only dimly aware of New Caledonia before writing this article, and had no idea that it was a mineral powerhouse.

RARE EARTHS

The metals known as the "rare earths" comprise three members of Group IIIB of the Periodic Table, scandium (Sc, 21), yttrium (Y, 39), and lanthanum (La, 57), and the 14 lanthanides that are filling the 4f electron shell: cerium (Ce, 58), praseodymium (Pr, 59), neodymium (Nd, 60), promethium (Pm, 61), samarium (Sm, 62), europium (Eu, 63), gadolinium (Gd, 64), terbium (Tb, 65), dysprosium (Dy, 66), homium (Ho, 67), erbium (Er, 68), thulium (Tm, 69), ytterbium (Yb, 70) and lutetium (Lu, 71). For physical properties, electron configurations and such details, the reader is referred to the Periodic Table websites to which links are given on the Physics index page. In the sequel, I shall refer to the elements by their chemical symbols, saving a lot of typing. Also included in this page will be some information on thorium (Th, 90), which is not a rare earth, but is associated with them in several respects.

The rare earths are not, in fact, very rare. Ce is the 25th most abundant element in the earth's crust at 60 ppm, ahead of tin and lead, and only a little less abundant than zinc. Pm, however has a natural abundance of zero, because its longest-lived isotope, Pm^{147}, has a half-life of only 2.6 days, so it has to be made freshly by nuclear reactions when any is desired. Nevertheless, it has been isolated as a metal. It was first prepared in 1963 at Oak Ridge. After Bohuslar Branner (1855-1935) predicted an element 61 in 1902, which was confirmed missing by H. B. G. Moseley in 1913-1914, Professor Luigi Rolla (1881-1960) and his student Fernandes reported its discovery in 1924 and named it Florentium, symbol Fr. Professor B. Smith Hopkins (1873-1952) of the University of

Illinois, with his students Yntema and Harris, claimed its discovery in 1926 and named it illinium, Il, to fill the last hole in the periodic table. The discovery was not confirmed. There was an "Illinium Dinner" at the university that year. Of course, none of these elements existed in Florence, Illinois or anywhere else at the time, and American chemistry lost a claim to fame. Arguments over priority of discovery of elements are a disgusting reality, and the rare earths had their share.

The rare earths can be divided into two groups, the first containing Y, La, Ce, Pr, Nd and Sm (Yttrium group), and the second Eu, Gd, Tb, Dy, Ho, Er, Tm, Yb and Lu (Dysprosium group), with Sc standing off at the side. All the elements are now available as pure metals. Before the 1950's, few if any of the dysprosium group had been actually isolated as metals, and only Ce was at all common, usually as an alloy with La, Nd and Pr called *mischmetall.* This word will be found spelled as "mischmetal" in English, and is a common item of commerce with various uses. Prices of the yttrium group currently range from $350 to $540 per kg, of the dysprosium group from $400 to $7500, while scandium is a breathtaking $18,000 per kg. For comparison, gold now is worth $12,110 per kg, so we see that any rare earth is cheaper than gold, except for scandium. Silver is worth $158 per kg, nickel $8.38 and chromium $7.50. The rare earths are available at quite reasonable prices, considering the difficulties of their processing and purification, and are slowly acquiring important industrial uses.

All the rare earths are typical transition metals. They are silver-white on a fresh surface, tarnish rapidly, are soft and usually ductile, and burn when heated. Ce can be cut with a knife, and can take fire if scratched. There are, of course, differences between them, but this is the general rule. None are of any use as structural metals. They are employed mainly in alloys and as compounds. Their chemical behaviour is so similar that they are exceedingly difficult to separate.

The sulphates of the yttrium group are soluble in water, while those of the dysprosium group are insoluble. The hydroxides of Sc, Y and La are the most basic, and the basicity decreases from Ce to Lu, as does the ionic radius. From Ce to Lu, the 4f electron shell is being filled. The 4f levels have about the same energy as the 6s levels, which are filled in the ground state. The often-heard explanation for the similarity of the rare earths based on the 4f levels being "hidden" is invalid. The ionic radius decreases from 0.106 nm for La^{+++} to 0.085 nm for Lu^{+++}, and is responsible for the slight but steady variation of properties along the series. Gd, with the electron configuration $Xe4f^75d6s^2$, has eight electrons with parallel spins, and so is highly paramagnetic. It is used in cryogenics for adabatic cooling. The valence of all the rare earths is most commonly +3. Ce, Pr and Tb also exhibit +4, while Sm, Eu, Tm and Yb show +2, in addition. Gd and Lu put one electron in a 5d level, encouraged by the stability of $4f^7$ and $4f^{14}$ configurations, respectively.

AUER VON WELSBACH

The name of Carl Auer, Freiherr von Welsbach (1858-1921), usually called Auer von Welsbach, is closely associated with the rare earths. He was born in Vienna, educated at the Technische Universitδt there, and later at Heidelberg, where he was a student of Robert Bunsen's, and received his PhD. At the age of 27, in 1885, he resolved the rare earth substance then known as didymium into two elements by laborious fractional crystallization of the ammonim nitrates. The part that gave greenish salts he named praseodymium (praseos is the leek in Greek, a green thing), while the part that gave pink salts he called neodymium. Neodymium is about twice as abundant as praseodymium. He was ennobled by Kaiser Franz Josef for his accomplishments.

Ce had been discovered by J. J. Berzelius and Hisinger in 1804, three years after the discovery of the first asteroid, Ceres, in 1801. Carl Gustav Mosander (1797-1858), Berzelius' assistant at Uppsala, prepared pure Ce in 1827, and in 1839 separated La from it. Then he separated a substance with a rose-colored oxide from La and called it didymium, from the Greek "didymos" for twin, as it was a twin to La. He discovered Y, Tb and Er in 1843. Four rare earths bear names derived from the village of Ytterby, near Stockholm, where the minerals containing them were found. These are yttrium, ytterbium, terbium and erbium. For a long time, the rare earths were a Swedish monopoly. By 1875, Ce, La and didymium were being made by the electrolysis of molten halides. In 1879, Paul Imile Lecoq de Boisbaudran (1838-1912) found Sm, and Dy in 1886. He was a spectroscopist who labored for 15 years to find gallium in 1875. Per Theodore Cleve (1840-1905), Svante Arrhenius' research advisor, isolated the much rarer and harder-to-separate Ho and Tm in 1879. Ho was christened after the Latin name of Stockholm, Tm after Thule, the "ultimate." Tm did not remain the ultimate for long. The periodic table, as then known (it appeared in 1869), gave little hint as to how many rare earths were to be expected. It was not clear that there were 14 lanthanides until the electronic basis of the periodic table was elucidated early in the 20th century. P. T. Cleve is also known for his work in diatoms, including the classification of *pseudonitzschia delicatissima*, which secretes the powerful toxin domoic acid that causes the rare amnesiac shellfish poisoning (ASP).

Shortly after Auer von Welsbach's separation of Pr and Nd, Lecoq de Boisboudran announced Dy, and Marignac Gd, both relatively common members of the dysprosium group, in 1886. All these new elements came from samples of the previously-known rare earths, as separation methods were discovered. Eu was found by Demarcay in 1900, and purified by G. Urbain and Lacombe in 1904. Its salts are pink, like those of Nd. Most salts of the rare earths are colorless. Auer von Welsbach had meanwhile been working in spectroscopy, and suspected that the substance isolated earlier by Marignac called ytterbium contained two elements. He succeeded in separating the elements in 1905, calling them aldebaranium and cassiopeium. Unfortunately, he was a few weeks later than G. C. Urbain, who had also separated the elements, and because of his priority was allowed to name them ytterbium and lutetium. These were the last two rare earths to be discovered, although by 1920 the periodic table showed that there was still one missing, a hole that was soon, but erroneously, plugged in Illinois.

Auer von Welsbach did not work only in the rare earths, though they were a favourite of his. His most widely known invention, the incandescent mantle, was patented in 1885, and called the "Auerlicht." In German, the incandescent mantle is known charmingly as the "Glóhstrumpf" or "glow-stocking." A knitted sleeve was soaked in a solution of nitrate salts and dried. To use it, it was first mounted on the lamp burner above the frame, and then lighted with a match. The organic parts burned, leaving fragile network of oxides in the path of the flame. When the !amp flame was lighted, and adjusted to be blue and nonluminous (the Argand burner was ideal for this) the mantle was heated to incandescence, giving a bright white light. The original mantles used Mg, La and Y. Pr, whose oxide is light brown, was used for the label printed on the mantle, which could be seen when the mantle was in use. They were very fragile, and so had a short life, and gave a greenish light.

A factory was established in Atzgersdorf in 1887 to make mantles, but it soon closed because of the deficiencies of the product. Auer von Welsbach worked assiduously to improve the mantle, and in 1890 discovered that thorium was better then magnesium, and in 1891 found the combination of 99% Th and 1% Ce that gave a long-lasting mantle with a brilliant white light. The factory reopened, and a new, larger factory was established in Althofen, in Carinthia (Kδrnten), where he made his home. The Wesbach Mantle soon spread world-wide, and revolutionized gas lighting. Illuminating gas no longer had to be carburetted to make luminous flames, and mantles worked excellently even with natural gas. Welsbach mantles are still used on camp lanterns and similar illuminating devices, though Y has replaced Th because of ignorant radiophobia.

The Welsbach and General Gas Mantle Company made mantles in Camden, NJ from the 1890's until 1941. More than 50 years later, this factory, then converted to other uses, was identified as a radiation hazard by the EPA. It is not possible to determine the facts of the case from published EPA documents, since they never give any survey figures or even say what their standards are, but it is very probable that this was an extremely minor hazard, giving gamma radiation levels scarcely higher than the cosmic-ray background.

It is reasonable that some thorium contamination was left in the area, but thorium is extremely feebly radioactive, and thoron contamination would have been equally negligible, easily counteracted by ventilation. The fear of traces of radioactivity has also driven thorium out of a reasonable use as an alloying agent for magnesium in aerospace alloys, where it was an equally small hazard. Fortunately, Y substitutes for Th in mantles, and a mixture of Y, Zr and other lanthanides is a substitute in Mg alloys. However, I have not seen any credible evidence that Th has ever been a health hazard because of its feeble radioactivity. There are far greater hazards in other places, and ignorant campaigns like this interfere with public protection rather than benefiting it by absorbing valuable resources.

Gas mantles give a light in every way competitive with electric light, and even cheaper. However, the bother of dealing with gas and mantles has made gas lighting disappear even in street lighting, where mantles lasted until late in the 20th century. Of course, the competition of electricity was realized from the first, since Edison's carbon-filament lamps were already widespread when the mantle was invented. Auer von Welsbach developed metal-filament lamps, patenting the *Auer-Oslicht* in 1902, which had an osmium filament made with powder metallurgy, in which Auer von Welsbach was a pioneer. This was a very good lamp, more economical than the Edison lamp, and a good deal brighter. It led in a few years to the replacement of the carbon filament with metallic filaments, of osmium, tantalum and finally tungsten. The "os" syllable remains in commercial names such as "Osram."

The observation that Ce easily caught fire, and burning sparks were easily made by striking it with a knife, led Auer von Welsbach to investigate "funkengebenden metalle" as an easy way to ignite a gas flame. In 1903 he patented an alloy of 70% Ce and 30% Fe which gave copious sparks when scratched by steel. In 1907, he founded a factory at Treibach for the manufacture of this *Cer-Eisen* to be used in lighters. The small pieces of alloy are called "flints" in English, "Zóndsteine" in German, and "piedras de encendedor" in Spanish, and are still widely used. A small cylinder of the alloy is held by spring pressure against a serrated steel wheel.

When the wheel is rotated by the thumb, a shower of sparks is produced. Pure Ce is not required. The cheaper alloy called "mischmetall" is just as good. This alloy is often incorrectly but commonly called "pyrophoric," a word that should be used for substances that take fire spontaneously, but there appears to be no existing term for *funkengebende* substances.

China is today the largest supplier of mischmetall. Two typical alloys are Ce 51, La 26, Nd 17, Pr 6 and Ce 53, Yb 24, Nd 16, Pr 5, Gd 2. The alloys do not have to be specially made. They result from the electrolysis of whatever combination happens to be in the natural ore, saving the cost of purification, which is totally unnecessary. When only the general properties of the rare earths is required, mischmetall is an economical alternative.

Mineralogy of the Rare Earths

Minerals containing rare earths are not especially rare, and the recent demand for rare earths has led to the discovery of new ores. There is little danger of a rare earth shortage at the present. The combination of potassium, rare earths, and phosphorus seems to be the last fraction that crystallizes from a magma, and is called KREEP by geologists. All the rare earths occur together in varying amounts, since nature is no better than chemists in separating them.

The principal ore of the rare earths has long been monazite, a dark, hard, heavy mineral that is found mainly in placer deposits, though vein deposits are known. It is found in commercial amounts in the U.S., Madagascar, Brazil, India, Sri Lanka and Australia. Monazite is principally $CePO_4$, cerous phosphate, but almost always contains a considerable amount of Th, together with La and other rare earths. Most Th, in fact, has been obtained as a by-product of the extraction of rare earths from monazite. Xenotime, YPO_4, found near Ytterby and the historical source of many rare earths, is similar. Euxenite is a mixed oxide of Y, Ce, Er and Nb, while Allanite is an aluminium-iron silicate with Y, Ce and Ca. Of these, only monazite is a commercial ore.

The improved techniques for production of rare earths that were developed after 1950, and the increasing demand, led to the discovery of a large resource of Bastnδsite at Mountain Pass, California. Bastnδsite is mainly $Ce_2(CO_3)_3$ or CeF_3 (?), a mixed fluoride-carbonate. The mineral is long known from Sweden, where its name comes from the village of Bastnδs. This deposit has made the U.S. one of the principal producers of rare earths, next to China and Russia.

Rare earths are now separated by ion exchange resins, which is much less tedious than the earlier fractional crystallization methods.

USES OF RARE EARTHS

Recently, the largest use of rare earths has been in glass polishing and ceramics, 39%, followed closely by automotive catalytic converters, 22%. 16% is used in permanent magnets, 12% for petroleum refining catalysts, 9% metallurgical uses, 1% in cathode-ray tube phosphors, and 1% in miscellaneous uses, including lighter flints. As an example of the metallurgical uses, mischmetall was used in the steel alloy for the Alaska crude oil pipline. The cathode ray tube phosphor is Eu-doped YVO_4 (yttrium vanadate), giving the red colour. Other phosphors also contain rare earths.

Didymium, the mixture of Pr and Nd that Auer von Welsbach separated in 1885, is used in glass for welders' and glass-workers' goggles. The violet-tinted glass is 80% transparent in the visible, but cuts off abruptly at about 340 nm for UV protection, and, remarkably, over a short range near 590 nm that blocks the Na glare. Neodymium-yttrium-garnet (NdYAG) and gadolinium-yttrium garnet (GdYAG) are laser materials, alternatives to ruby.

Rare earths are used in the permanent magnet alloys samarium-cobalt and neodymium-iron-boron. Sm-Co appeared in the late 1960's, Nd-Fe-B in 1983. Sm-Co magnets have the composition $SmCo_5$ or Sm_2Co_{17}, and Nd-Fe-B magnets are $Nd_2Fe_{14}B$. The 2-17 alloy gives better temperature characteristics than the 1-5. These are very hard and difficult materials that must be shaped by sintering or compression bonded with a bonding agent, and worked with diamond tools. A typical Sm-Co material has a remanent flux of 10,700 gauss and a coercive force of 10,300 oersted, giving a BH product of 28 MG-oe.

A Grade 30 La-Fe-B powder has a remanent flux of 11,400 gauss and a coercive force of 10,400 oersted, for BH = 30 MG-oe. There are many different materials, but this shows typical high performance. 1-5 Sm-Co has a Curie temperature of 750°C, and 2-17 825°C, but the maximum service temperature is about 350°C. La-Fe-B has a lower Curie temperature, 310°C, and a maximum service temperature of 150°C.

The corrosion resistance of La-Fe-B is low, but that of Sm-Co is high. The mechanical strength of brittle Sm-Co is low, while that of La-Fe-B is moderate. The density of La-Fe-B is about 7.5 g/cc, and that of Sm-Co is 8.3 g/cc. Sm-Co is expensive because of the cobalt. These materials require a very high magnetizing field, which is usually applied after assembly. These crystals are actually antiferromagnetic, with the lanthanide and cobalt or iron moments opposite. However, they have very high magnetic anisotropy, which makes them good permanent magnets, with a very high internal coercive force.

The search for a light, cheap rechargable battery with high energy density continues, but the lead-acid cell is hard to beat. A recent contender has been the Ni-Cd rechargeable battery, and more recently the Ni-metal hydride (Ni/MH) battery, which avoids the environmental restrictions on cadmium. These batteries have used the expensive cobalt, however, and have a limited life due to expansion and contraction of the plates during the charge-discharge cycle. An alloy has been developed at Brookhaven National Laboratories that uses La, Ni and Sn, so that it is free of expensive Co and toxic Cd. It was found that putting slightly more Ni/Sn in than the usual 1 La to 5 Ni/Sn prevented the stressing of the plates on charge and discharge, making a much more durable battery.

Ce_2O_3 is the catalyst used in self-cleaning ovens, and possibly also in automotive catalytic converters that break down unburned hydrocarbons in engine exhaust.

SCANDIUM

Scandium, a light metal between calcium and titanium in the first long period of the periodic table, acts much more like a rare earth than either of these, or of boron, which it might also resemble. It was discovered by Lars Fredrik Nilson in 1876 (or 1878, or 1879, according to different sources) in scandinavian euxinite, and was recognized by Cleve as the "eka-boron" predicted by Mendeleev in his periodic table of a few years before. It is light, like aluminium, but has a much higher melting point, which suggests aerospace uses. However, its cost would seem to put it out of the competition, though it would be very similar to titanium in this application. Scandium reacts readily with water to displace H_2, which would seem to eliminate it as a structural material. Scandium salts are colorless, and it is not paramagnetic.

The USSR was the main producer of scandium, but ceased export of it in 1968, after which its price went sky-high, to some $75,000 per kg until other sources were found. It seems to have been important in some laser application, but just what is not evident, and information is lacking. Scandium is used in aluminium alloys for baseball bats, bicycle frames and handguns. It is also used in high-intensity metal halide lamps, as the oxide scandia, Sc_2O_3. ScI_3 is added to mercury vapour lamps to make the light more like sunlight. Sc^{46}, a beta-emitter with a half-life of 85 days, is used as a tracer in oil refineries. Natural scandium is 100% Sc^{45}, with a rather large cross-section for thermal neutron absorption. It is expected that scandium may be used in fuel cells.

Scandium is believed to be the trace ion that gives the

blue colour to beryl in the gem aquamarine. It is relatively much more common in the Sun than on earth, where it is 50th in order of abundance.

THORIUM

Thorium, atomic number 90, was first isolated by J. J. Berzelius in 1828 or 1829. The heavy (d = 11.72 g/cc), shiny white metal soon tarnishes in air, gradually becoming black from a layer of oxide. The pure oxide, thoria, ThO_2, however, forms clear cubic crystals with the high index of refraction of 2.2 and specifc gravity 10.03. It is usually encountered as a white powder. The mineral *thorianite*, $(Th,U)O_2$, is black, with hardness Mohs 6.5 and specific gravity 8.0-9.7. The metal is face-centered cubic, soft and ductile. Thoria melts at 3300°C, nearly as high as tungsten does, which makes it an excellent refractory, used for laboratory crucibles. Thoria, up to 30%, also makes optical glass of a high index of refraction. Thorium hardens magnesium without interfering with its passivity, giving a strong, light alloy with aerospace uses. We have already noted its use as the principal component of incandescent mantles, although Y is a replacement. Thoria also is a catalyst in the oxidation of NH_3 to HNO_3, and other reactions.

Powdered thorium is pyrophoric (in the proper sense of the word), like many powdered metals (Al, Mg for example). It is attacked only slowly by water, but dissolves readily in hydrochloric acid. Only hot sulphuric acid attacks thoria. Its hydroxide, $Th(OH)_4$, makes a gelatinous precipitate, and is amphoteric, dissolving in alkalis. Although thorium occurs in minerals where it is the principal component, as in thorianite, ThO_2, it is commonly associated with the rare earths, and has been obtained mainly as a by-product in their extraction, and has benefited from the improvements in rare-earth metallurgy. Thorium occupies the same place at the head of the actinide series in the periodic table that cerium occupies in the lanthanide series.

Thorium is so rare that its chemical toxicity is not an issue, and no specific biological hazards are known. It is feebly radioactive, and widely distributed in igneous rocks. Basalt shows a thorium activity of 270-405 pCi/kg, and granite 1890 pCi/kg. There is little thorium in sedimentary rocks. In crustal abundance, it is between Pb and Be, the 34th in abundance, but has not been concentrated as well as these elements into economic ores. The Nuclear Regulatory Commission exempts items containing less than 2g or 0.25% of thorium from control, as well as optical glass containing less than 30% Th by weight.

The only naturally-occurring isotope of Th is Th^{232}, an alpha-emitter with a half-life of 1.39×10^{10} years. Tiny amounts of radioactive isotopes of Th occur in the decay chains of U^{238} and U^{235}. By the absorption of a neutron, Th^{232} becomes the fissile isotope Th^{233}, which can be used in nuclear reactors. Advantage has not been taken of this possibility, however. The end of the thorium radioactive series is Pb^{208}, so Th^{232}/Pb^{208} ratios can be used for geological dating. One member of the decay chain is the gas Rn^{220}, also called thoron, which is a hazard similar to the radon that occurs in the uranium series. However, it is very rare, since it has the short half-life of 54.5s, compared to 3.825 days for radon, and can scarcely be a hazard under normal circumstances (for equal thorium and uranium activities, thoron will be 0.00016 times less concentrated than radon).

It's good to remember that all nuclides with $Z > 83$ are radioactive, and even Bi^{209}, the only naturally-occurring isotope of bismuth, has a half-life of about 3×10^{17} years, so, strictly speaking, it is not stable either. Only the nuclides with half-lives of about 10^9 years or more have survived since the origin of the earth, and there are only three of these, Th^{232}, U^{235} and U^{238}. All the other elements in this region of the periodic table are either contained in the radioactive chains headed by these nuclides, or are artificially made. Of these, only Bi^{210} and three isotopes of uranium have long half-lives, which means that they will exist in reasonable amounts. The difficulty in isolating radium, with a half-life of 1620 years, is evidence of their rarity.

K^{40}, a rare isotope of potassium that is radioactive, is present in the earth's crust at 0.0029%, while the abundance of Th^{232} is 0.00115%. Even allowing for the radioactive daughters of the thorium, which potassium does not have, and the 10-times longer half-life of the thorium, natural gamma radioactivity due to potassium is larger than that due to thorium. I hope the EPA does not take fright at this and try to remove potassium from the environment.

Thermionic cathodes used in vacuum tubes emit electrons when heated. A tungsten filament heated to 2500K (white heat) is a useful emitter, but not a copious one. It was found in 1904 that barium and strontium oxide surfaces were copious emitters when heated only to 1000K (red heat), but were easily destroyed by ion emission when used at higher plate voltages. In 1914, Langmuir and Rogers found that a good emitter could be made by adding 1-2% ThO_2 to the tungsten. The filament had only to be heated to 1900K (orange heat) in this case. Before use, the filament had to be heated briefly to 2100K to activate it. The thorium is adsorbed as a monoatomic layer at the surface of the tungsten, reducing the work function from 4.52V to 2.63V, making it easier for the electrons to get out. As the thorium boils off, more diffuses from the bulk of the filament to replace it. These "thoriated tungsten" filaments are still used in thermionic emitters, and were widely used in transmitting tubes for plate voltages up to 5000V.

Thorium is also found as a coating for tungsten welding rods, and in electric arc carbon cores.

Chapter 10

Vanadium

Vanadium is the chemical element with the symbol V and atomic number 23. It is a soft, silvery grey, ductile transition metal. The formation of an oxide layer stabilizes the metal against oxidation. Andrıs Manuel del Rνo discovered vanadium in 1801 by analyzing the mineral vanadinite, and named it *erythronium*. Four years later, however, he was convinced by other scientists that erythronium was identical to chromium. The element was rediscovered in 1831 by Nils Gabriel Sefstrφm, who named it vanadium after the Norse goddess of beauty, Vanadis (Freyja). Both names were attributed to the wide range of colors found in vanadium compounds.

The element occurs naturally in about 65 different minerals and in fossil fuel deposits. It is produced in China and Russia from steel smelter slag; other countries produce it either from the flue dust of heavy oil, or as a byproduct of uranium mining. It is mainly used to produce specialty steel alloys such as high speed tool steels. The compound vanadium pentoxide is used as a catalyst for the production of sulfuric acid. Vanadium is found in many organisms, and is used by some life forms as an active centre of enzymes.

CHARACTERISTICS

Vanadium is a soft, ductile, silver-grey metal. It has good resistance to corrosion and it is stable against alkalis, sulfuric and hydrochloric acids. It is oxidized in air at about 933 K (660°C), although an oxide layer forms even at room temperature.

Isotopes

Naturally occurring vanadium is composed of one stable isotope ^{51}V and one radioactive isotope ^{50}V. The latter has a half-life of 1.5×10^{17} years and a natural abundance 0.25%. ^{51}V has a nuclear spin of 7/2 which is useful for NMR spectroscopy. A number of 24 artificial radioisotopes have been characterized, ranging in mass number from 40 to 65.

The most stable of these isotopes are ^{49}V with a half-life of 330 days, and ^{48}V with a half-life of 15.9735 days. All of the remaining radioactive isotopes have half-lives shorter than an hour, most of which are below 10 seconds. At least 4 isotopes have metastable excited states. Electron capture is the main decay mode for isotopes lighter than the ^{51}V. For the heavier ones, the most common mode is beta decay. The electron capture reactions lead to the formation of element 22 (titanium) isotopes, while for beta decay, it leads to element 24 (chromium) isotopes.

Chemistry and Compounds

The chemistry of vanadium is noteworthy for the accessibility of four adjacent oxidation states. The common oxidation states of vanadium are +2 (lilac), +3 (green), +4 (blue) and +5 (yellow). Vanadium(II) compounds are reducing agents, and vanadium(V) compounds are oxidizing agents. Vanadium(IV) compounds often exist as vanadyl derivatives which contain the VO^{2+} centre.

Ammonium vanadate(V) (NH_4VO_3) can be successively reduced with elemental zinc to obtain the different colors of vanadium in these four oxidation states. Lower oxidation states occur in compounds such as $V(CO)_6$,$[V(CO)_6]^-$ and substituted derivatives.

The vanadium redox battery utilizes these oxidation states; conversions of these oxidation states is illustrated by the reduction of a strongly acidic solution of a vanadium(V) compound with zinc dust. The initial yellow colour characteristic of the vanadate ion, VO_4^{3-}, is replaced by the blue colour of $[VO(H_2O)_5]^{2+}$, followed by the green colour of $[V(H_2O)_6]^{3+}$ and then violet, due to $[V(H_2O)_6]^{2+}$.

The most commercially important compound is vanadium pentoxide, which is used as a catalyst for the production of sulfuric acid. This compound oxidizes sulfur dioxide (SO_2) to the trioxide (SO_3). In this redox reaction, sulfur is oxidized from +4 to +6, and vanadium is reduced from +5 to +3:

$$V_2O_5 + 2\ SO_2 \rightarrow V_2O_3 + 2\ SO_3$$

The catalyst is regenerated by oxidation with air:

$$V_2O_3 + O_2 \rightarrow V_2O_5$$

Oxy and Oxo Compounds

The oxyanion chemistry of vanadium(V) is complex. The vanadate ion, VO_4^{3-}, is present in dilute solutions at high pH. On acidification, HVO_4^{2-} and $H_2VO_4^-$ are formed, analogous to HPO_4^{2-} and $H_2PO_4^-$.

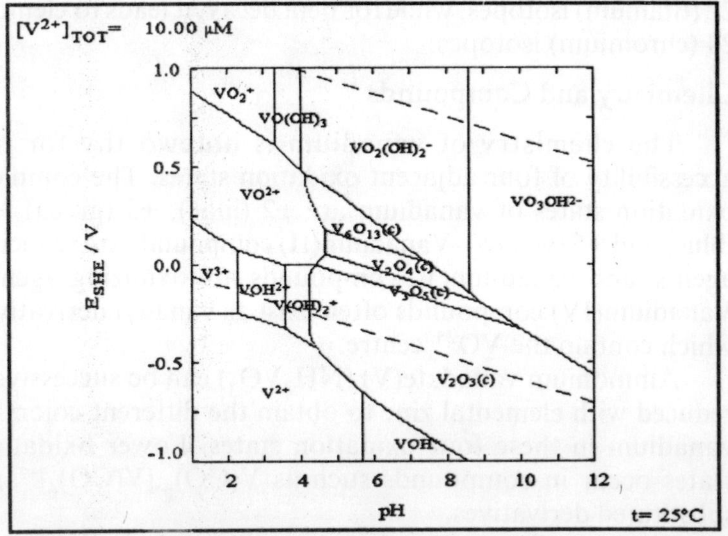

The acid dissociation constants for the vanadium and phosphorus series are remarkably similar. In more concentrated solutions many polyvanadates are formed. Chains, rings and clusters involving tetrahedral vanadium, analogous to the polyphosphates, are known.

In addition, clusters such as the decavanadates $V_{10}O_{28}^{4-}$ and $HV_{10}O_{28}^{3-}$, which predominate at pH 4-6, are formed in which compound is octahedral about vanadium. The correspondence between vanadate and phosphate chemistry can be attributed to the similarity in size and charge of phosphorus(V) and vanadium(V). Orthovanadate VO_4^{3-} is used in protein crystallography to study the biochemistry of phosphate.

Halide Compounds

Several halides are known for oxidation states +2, +3 and +4. VCl_4 is the most important commercially. This liquid is mainly used as a catalyst for polymerization of dienes.

Coordination Compounds

Vanadium's early position in the transition metal series lead to three rather unusual features of the coordination chemistry of vanadium. Firstly, metallic vanadium has the electronic configuration $[Ar]4s^2 3d^3$, so compounds of vanadium are relatively electron-poor. Consequently, most binary compounds are Lewis acids (electron pair acceptors); examples are all the halides forming octahedral adducts with the formula VX_nL_{6-n} (X = halide; L = other ligand).

Secondly, the vanadium ion is rather large and can achieve coordination numbers higher than 6, as is the case in $[V(CN)_7]^{4-}$. Thirdly, the vanadyl ion, VO^{2+}, is featured in many complexes of vanadium(IV) such as vanadyl acetylacetonate ($V(=O)(acac)_2$). In this complex, the vanadium is 5-coordinate, square pyramidal, meaning that a sixth ligand, such as pyridine, may be attached, though the association constant of this process is small. Many 5-coordinate vanadyl complexes have a trigonal bypyramidal geometry, such as $VO(Cl_2(NMe_3)_2$.

Organometallic Compounds

Organometallic chemistry of vanadium is well developed, but organometallic compounds are of minor commercial significance. Vanadocene dichloride is a versatile starting reagent and even finds minor applications in organic chemistry. Vanadium carbonyl, $V(CO)_6$, is a rare example of a metal carbonyl containing an unpaired electron, but which exists without dimerization. The addition of an electron yields $V(CO)_6^-$ (isoelectronic with $Cr(CO)_6$), which may be further reduced with sodium in liquid ammonia to yield $V(CO)_5^{3-}$ (isoelectronic with $Fe(CO)_5$)

VANADIUM(V) OXIDE AS A CATALYST IN THE CONTACT PROCESS

The Overall Reaction

During the Contact Process for manufacturing sulphuric acid, sulphur dioxide has to be converted into sulphur trioxide. This is done by passing sulphur dioxide and oxygen over a solid vanadium(V) oxide catalyst.

$$SO_2 + \frac{1}{2}O_2 \xrightarrow{V_2O_5} SO_3$$

How Reaction Works

This is a good example of the ability of transition metals and their compounds to act as catalysts because of their ability to change their oxidation state (oxidation number).

The sulphur dioxide is oxidised to sulphur trioxide by the vanadium(V) oxide. In the process, the vanadium(V) oxide is reduced to vanadium(IV) oxide.

$$SO_2 + V_2O_5 \longrightarrow SO_3 + V_2O_4$$

The vanadium(IV) oxide is then re-oxidised by the oxygen.

$$V_2O_4 + \frac{1}{2}O_2 \longrightarrow V_2O_5$$

Although the catalyst has been temporarily changed during the reaction, at the end it is chemically the same as it started.

Vanadium's Oxidation States

Vanadium has oxidation states in its compounds of +5, +4, +3 and +2. This section looks at ways of changing between them. It starts with a bit of description, and then goes on to look at the reactions in terms of standard redox potentials (standard electrode potentials).

Reducing Vanadium(V) in Stages to Vanadium(II)

The usual source of vanadium in the +5 oxidation state is ammonium metavanadate, NH_4VO_3. This isn't very soluble in water and is usually first dissolved in sodium hydroxide solution. The solution can be reduced using zinc and an acid - either hydrochloric acid or sulphuric acid, usually using moderately concentrated acid. The exact vanadium ion present in the solution is very complicated, and varies with the pH of the solution. The reaction is done under acidic conditions when the main ion present is VO_2^+ - called the dioxovanadium(V) ion. If you do the reaction in a small flask, it is normally stoppered with some cotton wool. This allows hydrogen (produced from a side reaction between the zinc and acid) to escape. At the same time it stops much air from entering. This prevents re-oxidation of the lower oxidation states of vanadium (particularly the +2 state) by oxygen in the air.

The reaction is usually warmed so that the changes happen in a reasonable time.

The reduction is shown in two stages. Some individual important colours are shown, but the process is one continuous change from start to finish.

The reduction from +5 to +4

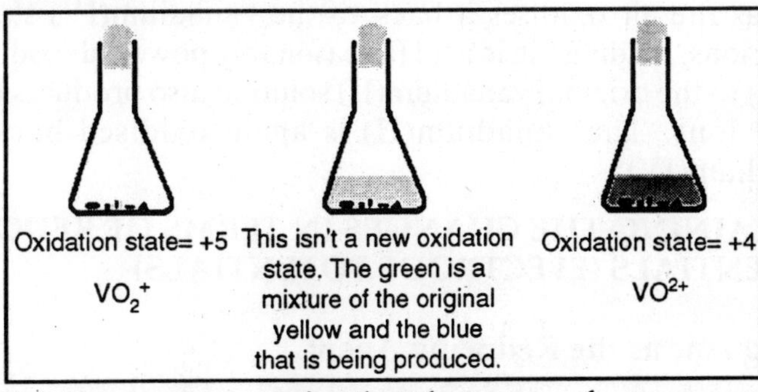

It is important to notice that the green colour you see isn't actually another oxidation state. it is just a mixture of the original yellow of the +5 state and the blue of the +4. Be very careful with the formulae of the two vanadium ions - they are very easy to confuse!

The reduction from +4 to +2

The colour changes just continue.

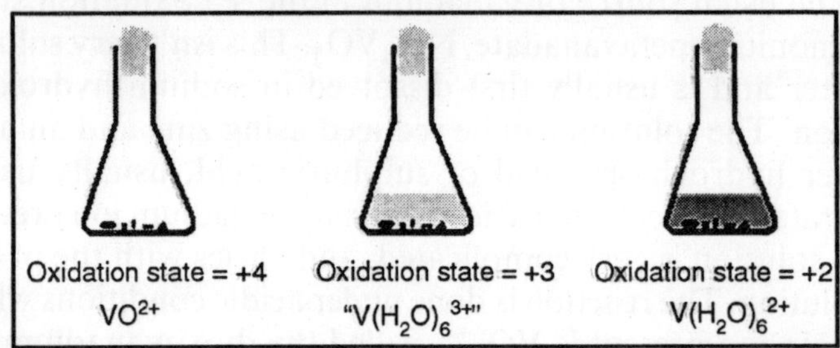

The reason for the inverted commas around the vanadium(III) ion is that this is almost certainly a simplification. The exact nature of the complex ion will depend on which acid you use in the reduction process. The simplification is probably reasonable at this level.

Re-Oxidation of the Vanadium(II)

The vanadium(II) ion is very easily oxidised. If you remove the cotton wool from the flask and pour some solution into a test tube, it turns green because of its contact with oxygen in the air. It is oxidised back to vanadium(III). If it is allowed to stand for a long time, the solution eventually turns blue as the air oxidises it back to the vanadium(IV) state - VO^{2+} ions. Adding nitric acid (a reasonably powerful oxidising agent) to the original vanadium(II) solution also produces blue VO^{2+} ions. The vanadium(II) is again oxidised back to vanadium(IV).

EXPLAINING THE CHANGES IN TERMS OF REDOX POTENTIALS (ELECTRODE POTENTIALS)

Using zinc as the Reducing Agent

The first stage of the series of reductions

Let's look at the first stage of the reduction - from VO_2^+

to VO^{2+}. The redox potential for the vanadium half-reaction is given by:

$$VO_2{}^{+}{}_{(aq)} + 2H^{+}{}_{(aq)} + e^{-} \rightleftharpoons VO^{2+}{}_{(aq)} + H_2O_{(l)}$$

$$E^{o} = +1.00v$$

The corresponding equilibrium for the zinc is:

$$Zn^{2+}{}_{(aq)} + 2e^{-} \rightleftharpoons Zn_{(s)} \qquad E^{o} = -0.76v$$

The simple principle is that if you couple two of these half-reactions together, the one with the more positive E° value will move to the right; the one with the more negative (or less positive) E° moves to the left.

So . . . if you mix together zinc and $VO_2{}^{+}$ ions in the presence of acid to provide the H^{+} ions:

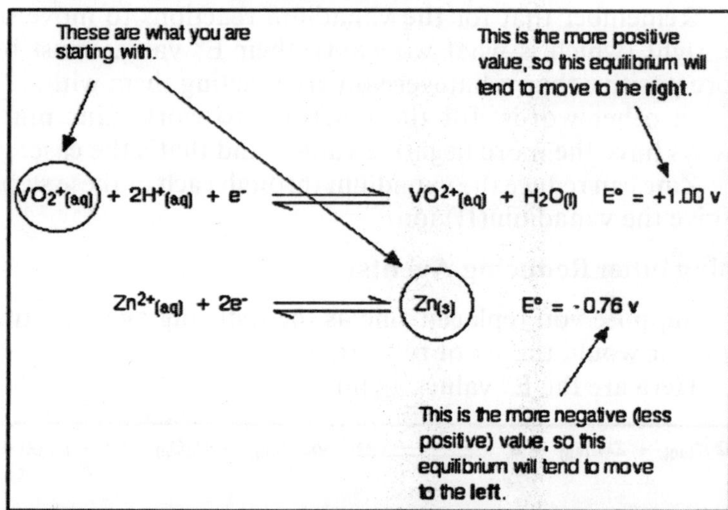

That converts the two equilibria into two one-way reactions. You can write these down and combine them to give the ionic equation for the reaction if you want to.

$$2 \times (VO_2{}^{+}{}_{(aq)} + 2H^{+}{}_{(aq)} + e^{-} \longrightarrow VO^{2+}{}_{(aq)} + H_2O_{(l)})$$

$$Zn_{(s)} \longrightarrow Zn^{2+}{}_{(aq)} + 2e^{-}$$

$$2VO_2{}^{+}{}_{(aq)} + 4H^{+}{}_{(aq)} + Zn_{(s)} \longrightarrow 2VO^{2+}{}_{(aq)} + 2H_2O_{(l)} + Zn^{2+}{}_{(aq)}$$

The other stages of the reaction

Here are the E° values for all the steps of the reduction from vanadium(V) to vanadium(II):

$$VO_2^+{}_{(aq)} + 2H^+{}_{(aq)} + e^- \rightleftharpoons VO^{2+}{}_{(aq)} + H_2O_{(l)} \quad E^\circ = +1.00\,v$$

$$VO^{2+}{}_{(aq)} + 2H^+{}_{(aq)} + e^- \rightleftharpoons V^{3+}{}_{(aq)} + H_2O_{(l)} \quad E^\circ = +0.34\,v$$

$$V^{3+}{}_{(aq)} + e^- \rightleftharpoons V^{2+}{}_{(aq)} \quad E^\circ = -0.26\,v$$

. . . and here is the zinc value again:

$$Zn^{2+}{}_{(aq)} + 2e^- \rightleftharpoons Zn_{(s)} \qquad E^o = -0.76v$$

Remember that for the vanadium reactions to move to the right (which is what we want), their E° values must be more positive than whatever you are reacting them with.

In other words, for the reactions to work, zinc must always have the more negative value - and that's the case.

Zinc can reduce the vanadium through each of these steps to give the vanadium(II) ion.

Using other Reducing Agents

Suppose you replaced zinc as the reducing agent by tin. How far would the set of reductions go this time?

Here are the E° values again:

$$VO_2^+{}_{(aq)} + 2H^+{}_{(aq)} + e^- \rightleftharpoons VO^{2+}{}_{(aq)} + H_2O_{(l)} \quad E^\circ = +1.00\,v$$

$$VO^{2+}{}_{(aq)} + 2H^+{}_{(aq)} + e^- \rightleftharpoons V^{3+}{}_{(aq)} + H_2O_{(l)} \quad E^\circ = +0.34\,v$$

$$V^{3+}{}_{(aq)} + e^- \rightleftharpoons V^{2+}{}_{(aq)} \quad E^\circ = -0.26\,v$$

. . . and here is the tin value:

$$Sn^{2+}{}_{(aq)} + 2e^- \rightleftharpoons Sn_{(s)} \qquad E^o = -0.14v$$

In order for each reduction to happen, the vanadium reaction has to have the more positive E° value because we want it to go to the right. That means that the tin must have the more negative value.

In the first vanadium equation (from +5 to +4), the tin value is more negative.

That works OK. In the second vanadium equation (from +4 to +3), the tin value is again the more negative.

That works as well.

But in the final vanadium reaction (from +3 to +2), tin no longer has the more negative E° value. Tin won't reduce vanadium(III) to vanadium(II).

Re-Oxidation of the vanadium(II)

The vanadium(II) oxidation state is easily oxidised back to vanadium(III) - or even higher.

Oxidation by hydrogen ions

You will remember that the original reduction we talked about was carried out using zinc and an acid in a flask stoppered with a piece of cotton wool to keep the air out.

Air will rapidly oxidise the vanadium(II) ions - but so also will the hydrogen ions present in the solution!

The vanadium(II) solution is only stable as long as you keep the air out, and in the presence of the zinc.

The zinc is necessary to keep the vanadium reduced. What happens if the zinc isn't there?

Look at these E° values:

$$V^{3+}_{(aq)} + e^- \rightleftharpoons V^{2+}_{(aq)} \qquad E^o = -0.26v$$

$$2H^+_{(aq)} + 2e^- \rightleftharpoons H_{2(q)} \qquad E^o = 0v$$

The reaction with the more negative E° value goes to the left; the reaction with the more positive (or less negative) one to the right.

That means that the vanadium(II) ions will be oxidised to vanadium(III) ions, and the hydrogen ions reduced to hydrogen.

Will the oxidation go any further - for example, to the vanadium(IV) state?

Have a look at the E° values and decide:

$$VO^{2+}{}_{(aq)} + 2H^{+}{}_{(aq)} + e^{-} \rightleftharpoons V^{3+}{}_{(aq)} + H_2O_{(l)}$$

$E^{o} = +0.34v$

$$2H^{+}{}_{(aq)} + 2e^{-} \rightleftharpoons H_{2(q)} \qquad E^{o} = 0v$$

In order for the vanadium equilibrium to move to the left, it would have to have the more negative E° value. It *hasn't* got the more negative E° value and so the reaction doesn't happen.

Oxidation by Nitric Acid

In a similar sort of way, you can work out how far nitric acid will oxidise the vanadium(II).

Here's the first step:

$$V^{3+}{}_{(aq)} + e^{-} \rightleftharpoons V^{2+}{}_{(aq)}$$

$E^{o} = -0.26v$

$$NO_3^{-}{}_{(aq)} + 4H^{+}{}_{(aq)} + 3e^{-} \rightleftharpoons NO_{(q)} + 2H_2O_{(l)}$$

$E^{o} = 0v$

The vanadium reaction has the more negative E° value and so will move to the left; the nitric acid reaction moves to the right.

Nitric acid will oxidise vanadium(II) to vanadium(III).

The second stage involves these E° values:

$$VO^{2+}{}_{(aq)} + 2H^{+}{}_{(aq)} + e^{-} \rightleftharpoons V^{3+}{}_{(aq)} + H_2O_{(l)}$$

$E^{o} = +0.34v$

$$NO_3^{-}{}_{(aq)} + 4H^{+}{}_{(aq)} + 3e^{-} \rightleftharpoons NO_{(q)} + 2H_2O_{(l)}$$

$E^{o} = +0.96v$

The nitric acid again has the more positive E° value and so moves to the right. The more negative (less positive) vanadium reaction moves to the left.

Nitric acid will certainly oxidise vanadium(III) to vanadium(IV).

Will it go all the way to vanadium(V)?

$$VO_{2(aq)}^{+} + 2H^{+}_{(aq)} + e^{-} \rightleftharpoons VO^{2+}_{(aq)} + H_2O_{(l)}$$

$$E^o = +1.00v$$

$$NO_3^{-}{}_{(aq)} + 4H^{+}_{(aq)} + 3e^{-} \rightleftharpoons NO_{(q)} + 2H_2O_{(l)}$$

$$E^o = +0.96v$$

No, it won't! For the vanadium reaction to move to the left to form the dioxovanadium(V) ion, it would have to have the more negative (less positive) E° value. It hasn't got a less positive value, and so the reaction doesn't happen.

OCCURRENCE

Metallic vanadium is not found in nature, but is known to exist in about 65 different minerals. Economically significant examples include patronite (VS_4), vanadinite ($Pb_5(VO_4)_3Cl$), and carnotite ($K_2(UO_2)_2(VO_4)_2{\cdot}3H_2O$). Much of the world's vanadium production is sourced from vanadium-bearing magnetite found in ultramafic gabbro bodies. Vanadium is mined mostly in South Africa, north-western China, and eastern Russia. In 2007 these three countries mined more than 95□% of the 58,600 tonnes of produced vanadium.

Vanadium is also present in bauxite and in fossil fuel deposits such as crude oil, coal, oil shale and tar sands. In crude oil, concentrations up to 1200□ppm have been reported. When such oil products are burned, the traces of vanadium may initiate corrosion in motors and boilers. An estimated 110,000 tonnes of vanadium per year are released into the atmosphere by burning fossil fuels. Vanadium has also been detected spectroscopically in light from the Sun and some other stars.

PRODUCTION

Most vanadium is used as ferrovanadium as an additive to improve steels. Ferrovanadium is produced directly by reducing a mixture of vanadium oxide, iron oxides and iron in an electric furnace. Vanadium-bearing magnetite iron ore is the main source for the production of vanadium. The vanadium ends up in pig iron produced from vanadium bearing magnetite. During steel production, oxygen is blown into the pig iron, oxidizing the carbon and most of the other impurities, forming slag. Depending on the used ore, the slag contains up to 25% of vanadium.

Vanadium metal is obtained via a multistep process that begins with the roasting of crushed ore with NaCl or Na_2CO_3 at about 850□°C to give sodium metavanadate ($NaVO_3$). An aqueous extract of this solid is acidified to give "red cake", a polyvanadate salt, which is reduced with calcium metal. As an alternative for small scale production, vanadium pentoxide is reduced with hydrogen or magnesium.

Many other methods are also in use, in all of which vanadium is produced as a byproduct of other processes.[21] Purification of vanadium is possible by the crystal bar process developed by Anton Eduard van Arkel and Jan Hendrik de Boer in 1925. It involves the formation of the metal iodide, in this example vanadium(III) iodide, and the subsequent decomposition to yield pure metal.

$$2\,V + 3\,I_2\,2 \rightleftharpoons VI_3$$

APPLICATIONS

Alloys

Approximately 85% of vanadium produced is used as ferrovanadium or as a steel additive. The considerable increase of strength in steel containing small amounts of vanadium was discovered in the beginning of 20th century, and from that time vanadium steel was used for applications in axles, bicycle frames, crankshafts, gears, and other critical components.

Vanadium forms stable nitrides and carbides, resulting in a significant increase in the strength of the steel. There are two groups of vanadium containing steel alloy groups. Vanadium high-carbon steel alloys containing 0.15 to 0.25 percent vanadium and high speed tool steels (HSS) with a vanadium content ranges from 1% to 5%. For high speed tool steels, a hardness above HRC 60 can be achieved. HSS steel is used in surgical instruments and tools.

Vanadium stabilizes the beta form of titanium and increases the strength and temperature stability of titanium. Mixed with aluminium in titanium alloys it is used in jet engines and high-speed airframes. One of the common alloys is Ti 6Al 4V a titanium alloy with 6% aluminium and 4% of vanadium.

Other Uses

Vanadium is compatible with iron and titanium, therefore vanadium foil is used in cladding titanium to steel. The moderate thermal neutron-capture cross-section and the short half-life of the isotopes produced by neutron capture makes vanadium a suitable material for the inner structure of a fusion reactor. Several vanadium alloys show superconducting behaviour.

The first A15 phase superconductor was a vanadium compound, V_3Si, which was discovered in 1952.Vanadium-gallium tape is used in superconducting magnets (17.5 teslas or 175,000 gauss). The structure of the superconducting A15 phase of V_3Ga is similar to that of the more common Nb_3Sn and Nb_3Ti.

The most common oxide of vanadium Vanadium pentoxide, V_2O_5, is used as a catalyst in manufacturing sulfuric acid by the contact process and as an oxidizer in maleic anhydride production.Vanadium pentoxide is also used in making ceramics. Another oxide of vanadium, vanadium dioxide VO_2, is used in the production of glass coatings, which blocks infrared radiation (and not visible light) at a specific temperature.Vanadium oxide can be used to induce colour centers in corundum to creat simulated alexandrite jewelry, although alexandrite in nature is a chrysoberyl.The possibility to use vanadium redox couples in both half-cells, thereby eliminating the problem of cross contamination by diffusion of ions across the membrane is the advantage of vanadium redox rechargeable batteries.

Vanadate can be used for protecting steel against rust and

corrosion by electrochemical conversion coating.Lithium vanadium oxide has been proposed for use as a high energy density anode for lithium ion batteries, at 745□Wh/l when paired with a lithium cobalt oxide cathode. It has ben proposed by some researchers that a small amount, 40 to 270□ppm, of vanadium in Wootz steel and Damascus steel, significantly improves the strength of the material, although it is unclear what the source of the vanadium was.

Chapter 11

Chromium

Chromium is a chemical element which has the symbol Cr and atomic number 24. It is a steely-gray, lustrous, hard metal that takes a high polish and has a high melting point. It is also odourless, tasteless, and malleable.

The name of the element is derived from the Greek word "chrτma" (χρωμα), meaning colour, because many of its compounds are intensely colored.

It was discovered by Louis Nicolas Vauquelin in the mineral crocoite (*lead chromate*) in 1797. Crocoite was used as a pigment, and after the discovery that the mineral chromite also contains chromium this latter mineral was used to produce pigments as well.

Chromium was regarded with great interest because of its high corrosion resistance and hardness. A major development was the discovery that steel could be made highly resistant to corrosion and discoloration by adding chromium and nickel to form stainless steel. This application and electroplating are currently the highest-volume uses of the metal.

Chromium and ferrochromium are produced from the single commercially viable ore, chromite, by silicothermic or aluminothermic reaction or by roasting and leaching processes. Although trivalent chromium (Cr(III)) is required in trace amounts for sugar and lipid metabolism in humans and its deficiency may cause a disease called chromium deficiency, hexavalent chromium (Cr(VI)) is a toxin and a carcinogen, so that abandoned chromium production sites need environmental cleanup.

Isotopes

Naturally occurring chromium is composed of three stable isotopes; ^{52}Cr, ^{53}Cr, and ^{54}Cr with ^{52}Cr being the most abundant (83.789% natural abundance). Nineteen radioisotopes have been characterized with the most stable being ^{50}Cr with a half-life of (more than) 1.8×10^{17} years, and ^{51}Cr with a half-life of 27.7 days. All of the remaining radioactive isotopes have half-lives that are less than 24 hours and the majority of these have half-lives that are less than 1 minute. This element also has 2 meta states.

^{53}Cr is the radiogenic decay product of ^{53}Mn. Chromium isotopic contents are typically combined with manganese isotopic contents and have found application in isotope geology.

Mn-Cr isotope ratios reinforce the evidence from ^{26}Al and ^{107}Pd for the early history of the solar system. Variations in $^{53}Cr/^{52}Cr$ and Mn/Cr ratios from several meteorites indicate an initial $^{53}Mn/^{55}Mn$ ratio that suggests Mn-Cr isotopic composition must result from in-situ decay of ^{53}Mn in differentiated planetary bodies. Hence ^{53}Cr provides additional evidence for nucleosynthetic processes immediately before coalescence of the solar system.

The isotopes of chromium range in atomic mass from 43□u (^{43}Cr) to 67□u (^{67}Cr). The primary decay mode before the most abundant stable isotope, ^{52}Cr, is electron capture and the primary mode after is beta decay.

REACTIONS OF CHROMIUM(III) IONS IN SOLUTION

The simplest ion that chromium forms in solution is the hexaaquachromium(III) ion - $[Cr(H_2O)_6]^{3+}$.

The acidity of the hexaaqua ions

In common with the other 3+ ions, the hexaaquachromium(III) ion is fairly acidic - with a pH for typical solutions in the 2 - 3 range.

The ion reacts with water molecules in the solution. A hydrogen ion is lost from one of the ligand water molecules:

$$\left[Cr(H_2O)_6\right]^{3+} + H_2O \rightleftharpoons \left[Cr(H_2O)_5(OH)\right]^{2+} + H_2O^+$$

The complex ion is acting as an acid by donating a hydrogen ion to water molecules in the solution. The water is, of course, acting as a base by accepting the hydrogen ion.

Because of the confusing presence of water from two different sources (the ligands and the solution), it is easier to simplify this:

$$\left[Cr(H_2O)_6\right]^{3+}(aq) \rightleftharpoons \left[Cr(H_2O)_5(OH)\right]^{2+}(aq) + H^+_{(aq)}$$

However, if you write it like this, remember that the hydrogen ion isn't just falling off the complex ion. It is being pulled off by a water molecule in the solution. Whenever you write "$H^+_{(aq)}$" what you really mean is a hydroxonium ion, H_3O^+.

Ligand Exchange Reactions Involving Chloride or Sulphate Ions

The hexaaquachromium(III) ion is a "difficult to describe" violet-blue-grey colour. However, when it is produced during a reaction in a test tube, it is often green.

We nearly always describe the green ion as being $Cr^{3+}_{(aq)}$ - implying the hexaaquachromium(III) ion. That's actually an over-simplification.

What happens is that one or more of the ligand water molecules get replaced by a negative ion in the solution - typically sulphate or chloride.

Replacement of the water by Sulphate ions[sh]

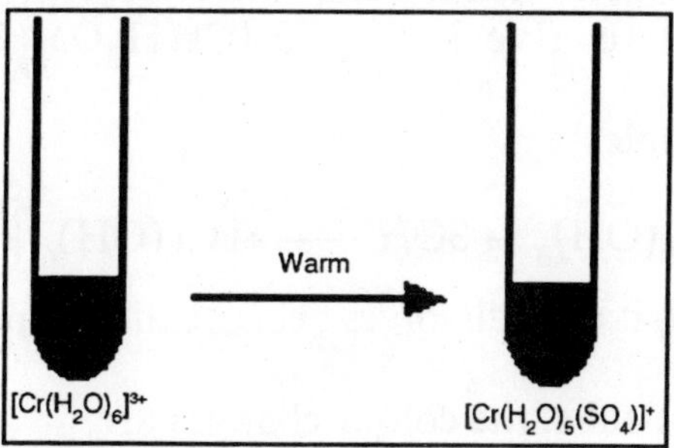

You can do this simply by warming some chromium(III) sulphate solution.

One of the water molecules is replaced by a sulphate ion. Notice the change in the charge on the ion.

Two of the positive charges are cancelled by the presence of the two negative charges on the sulphate ion.

Replacement of the Water by Chloride ions

In the presence of chloride ions (for example with chromium(III) chloride), the most commonly observed colour is green.

This happens when two of the water molecules are replaced by chloride ions to give the tetraaquadichlorochromium(III) ion - $[Cr(H_2O)_4Cl_2]^+$.

Once again, notice that replacing water molecules by chloride ions changes the charge on the ion.

REACTIONS OF HEXAAQUACHROMIUM(III) IONS WITH HYDROXIDE IONS

Hydroxide ions (from, say, sodium hydroxide solution) remove hydrogen ions from the water ligands attached to the chromium ion.

Once a hydrogen ion has been removed from three of the water molecules, you are left with a complex with no charge - a neutral complex. This is insoluble in water and a precipitate is formed.

$$\left[Cr(H_2O)_6\right]^{3+} + 3OH^- \longrightarrow \left[Cr(OH)_6\right]^{3+} 3H_2O$$

But the process doesn't stop there. More hydrogen ions are removed to give ions like $[Cr(H_2O)_2(OH)_4]^-$ and $[Cr(OH)_6]^{3-}$.

For example:

$$\left[Cr(H_2O)_3(OH)_3\right] + 3OH^- \longrightarrow \left[Cr(OH)_6\right]^{3-} + 3H_2O$$

The precipitate redissolves because these ions are soluble in water.

In the test-tube, the colour changes are:

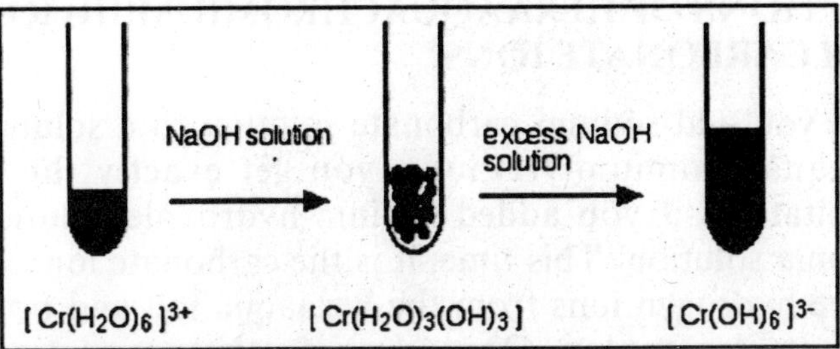

REACTIONS OF HEXAAQUACHROMIUM(III) IONS WITH AMMONIA SOLUTION

The ammonia acts as both a base and a ligand.

With a small amount of ammonia, hydrogen ions are pulled off the hexaaqua ion exactly as in the hydroxide ion case to give the same neutral complex.

$$\left[Cr(H_2O)_6\right]^{3+} + 3NH_3 \longrightarrow \left[Cr(H_2O)_3(OH)_3\right] + 3NH_4^+$$

That precipitate dissolves to some extent if you add an excess of ammonia (especially if it is concentrated).

The ammonia replaces water as a ligand to give hexaamminechromium(III) ions.

$$\left[Cr(H_2O)_6\right]^{3+} + 6NH_3 \rightleftharpoons \left[Cr(NH_3)_6\right]^{3} + 6H_2O$$

The colour changes are:

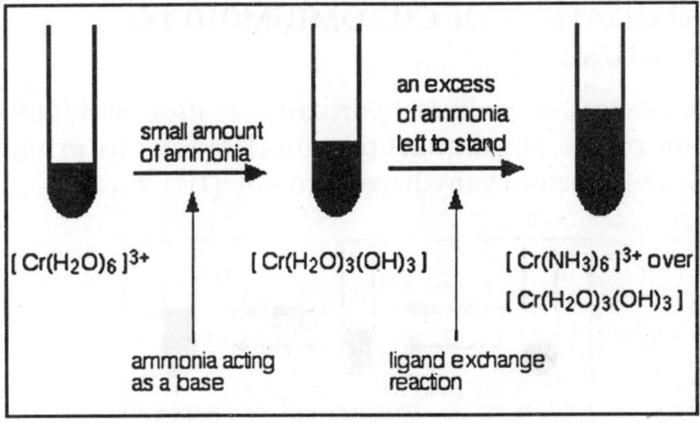

REACTIONS OF HEXAAQUACHROMIUM(III) IONS WITH CARBONATE IONS

If you add sodium carbonate solution to a solution of hexaaquachromium(III) ions, you get exactly the same precipitate as if you added sodium hydroxide solution or ammonia solution. This time, it is the carbonate ions which remove hydrogen ions from the hexaaqua ion and produce the neutral complex. Depending on the proportions of carbonate ions to hexaaqua ions, you will get either hydrogencarbonate ions formed or carbon dioxide gas from the reaction between the hydrogen ions and carbonate ions. The more usually quoted equation shows the formation of carbon dioxide.

$$2\left[Cr(H_2O)_6\right]^{3+} + 3CO_3^{2-} \longrightarrow 2$$

$$\left[Cr(H_2O)_3(OH)_3\right] + 3CO_2 + 3H_2O$$

Apart from the carbon dioxide, there is nothing new in this reaction:

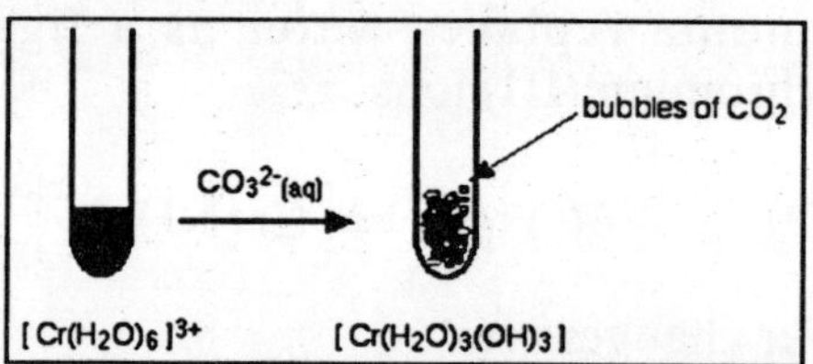

THE OXIDATION OF CHROMIUM(III) TO CHROMIUM(VI)

An excess of sodium hydroxide solution is added to a solution of the hexaaquachromium(III) ions to produce a solution of green hexahydroxochromate(III) ions.

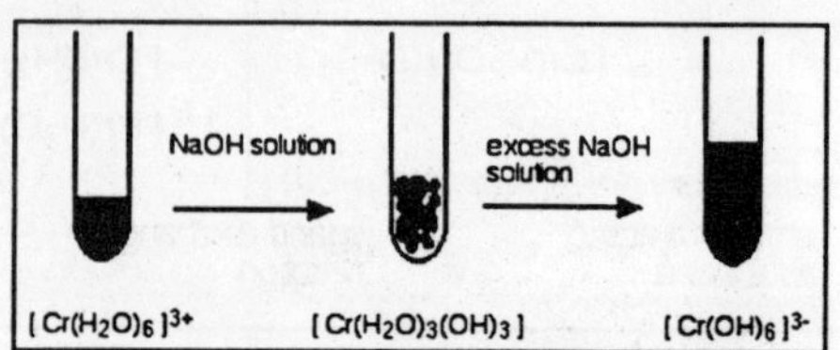

This is then oxidised by warming it with hydrogen peroxide solution. You eventually get a bright yellow solution containing chromate(VI) ions.

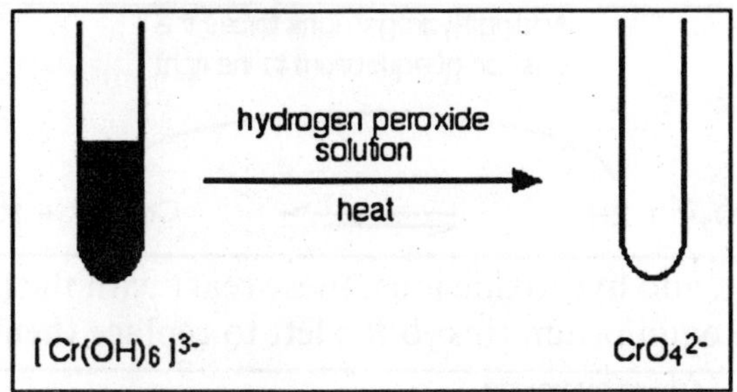

The equation for the oxidation stage is:

$$2\left[Cr(OH)_6\right]^{3+} + 3H_2O_2 \longrightarrow 2CrO_4^{2-} + 2OH^- + 8H_2O$$

CHROMIUM(VI) CHEMISTRY

The Chromate(VI)-Dichromate(VI) Equilibrium

You are probably more familiar with the orange dichromate(VI) ion, $Cr_2O_7^{2-}$, than the yellow chromate(VI) ion, CrO_4^{2-}.

Changing between them is easy:

If you add dilute sulphuric acid to the yellow solution it turns orange. If you add sodium hydroxide solution to the orange solution it turns yellow.

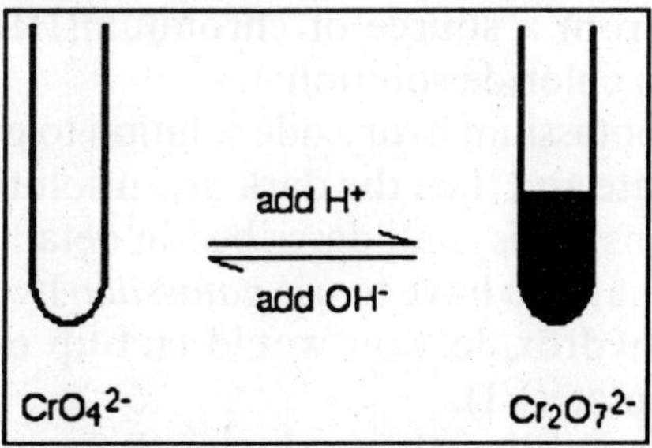

The equilibrium reaction at the heart of the interconversion is:

$$2CrO_4^{2-} + 2H^+ \rightleftharpoons Cr_2O_7^{2-} + H_2O$$

If you add extra hydrogen ions to this, the equilibrium shifts to the right. This is consistent with Le Chatelier's Principle.

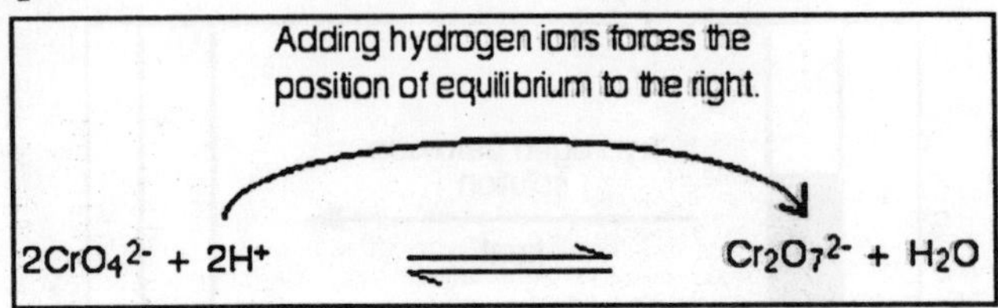

If you add hydroxide ions, these react with the hydrogen ions. The equilibrium tips to the left to replace them.

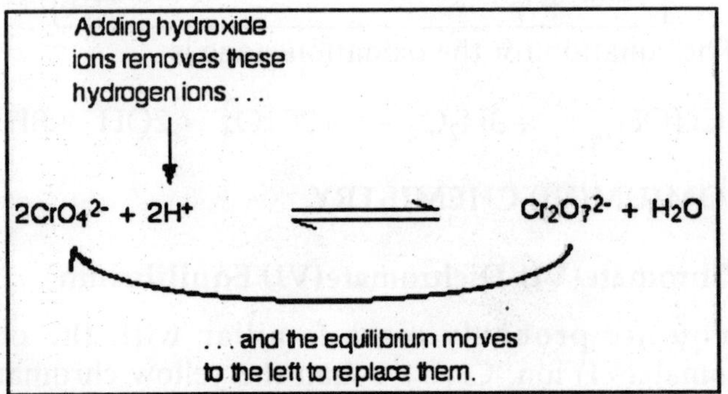

Making Potassium Dichromate(VI) Crystals

Potassium dichromate crystals can be made by a combination of the reactions we've already looked at on this page.

Starting from a source of chromium(III) ions such as chromium(III) chloride solution:

You add potassium hydroxide solution to give first a grey-green precipitate and then the dark green solution containing $[Cr(OH)_6]^{3-}$ ions. This is all described in detail further up the page. Notice that you have to use *potassium* hydroxide. If you used sodium hydroxide, you would end up eventually with *sodium* dichromate(VI).

Now you oxidise this solution by warming it with hydrogen peroxide solution. The solution turns yellow as potassium chromate(VI) is formed. This reaction is also described further up the page.

All that is left is to convert the yellow potassium chromate(VI) solution into orange potassium dichromate(VI) solution. You may remember that that is done by adding acid. This is described above if you have forgotten.

Unfortunately there is a problem here. Potassium dichromate will react with any excess hydrogen peroxide to give initially an unstable deep blue solution and it eventually gives the original chromium(III) ions again! To get around this, you first need to destroy any excess hydrogen peroxide.

This is done by boiling the solution. Hydrogen peroxide decomposes on heating to give water and oxygen. The solution is boiled until no more bubbles of oxygen are produced. The solution is heated further to concentrate it, and then concentrated ethanoic acid is added to acidify it. Orange crystals of potassium dichromate are formed on cooling.

The Reduction of Dichromate(VI) ions with zinc and an acid

Dichromate(VI) ions (for example, in potassium dichromate(VI) solution) can be reduced to chromium(III) ions and then to chromium(II) ions using zinc and either dilute sulphuric acid or hydrochloric acid.

Hydrogen is produced from a side reaction between the zinc and acid. This must be allowed to escape, but you need to keep air out of the reaction. Oxygen in the air rapidly re-oxidises chromium(II) to chromium(III).

An easy way of doing this is to put a bit of cotton wool in the top of the flask (or test-tube) that you are using. This allows the hydrogen to escape, but stops most of the air getting in against the flow of the hydrogen.

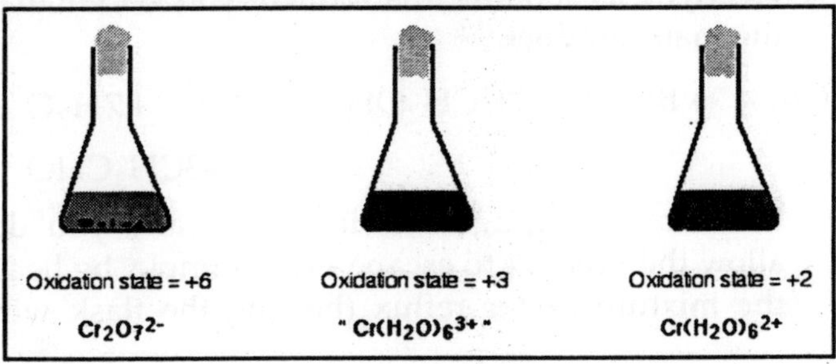

The reason for the inverted commas around the chromium(III) ion is that this is a simplification. The exact nature of the complex ion will depend on which acid you use in the reduction process. This has already been discussed towards the top of the page.

The equations for the two stages of the reaction are:

For the reduction from +6 to +3:

$$Cr_2O_7^{2-} + 14H^+ + 3Zn \longrightarrow 2Cr^{3+} + 7H_2O + 3Zn^{2+}$$

For the reduction from +3 to +2:

$$2Cr^{3+} + Zn \longrightarrow 2Cr^{2+} + Zn^{2+}$$

Using Potassium Dichromate(VI) as an Oxidising agent in Organic Chemistry

Potassium dichromate(VI) solution acidified with dilute sulphuric acid is commonly used as an oxidising agent in organic chemistry.

It is a reasonably strong oxidising agent without being so powerful that it takes the whole of the organic molecule to pieces! (Potassium manganate(VII) solution has some tendency to do that.)

It is used to:

- Oxidise secondary alcohols to ketones;
- oxidise primary alcohols to aldehydes;
- oxidise primary alcohols to carboxylic acids.

For example, with ethanol (a primary alcohol), you can get either ethanal (an aldehyde) or ethanoic acid (a carboxylic acid) depending on the conditions.

- If the alcohol is in excess, and you distil off the aldehyde as soon as it is formed, you get ethanal as the main product.

$$Cr_2O_7^{2-} + BH^+ + 3CH_3CH_2OH \longrightarrow 2Cr^{3+} + 7H_2O + 3CH_3CHO$$

- If the oxidising agent is in excess, and you don't allow the product to escape - for example, by heating the mixture under reflux (heating the flask with a condenser placed vertically in the neck) - you get ethanoic acid.

$$2Cr_2O_7^{2-} + 16H^+ + 3CH_3CH_2OH \longrightarrow 4Cr^{3+} + 11H_2O + 3CH_3COOH$$

In organic chemistry, these equations are often simplified to concentrate on what is happening to the organic molecules. For example, the last two could be written:

$$CH_3CH_2OH + [O] \longrightarrow CH_3CHO + H_2O$$

$$CH_3CH_2OH + 2[O] \longrightarrow CH_3COOH + H_2O$$

The oxygen written in square brackets just means "oxygen from an oxidising agent".

Using this same reaction to make chrome alum crystals

You will find chrome alum under all sorts of different names:

- Chrome alum
- Potassium chromium(III) sulphate
- Chromium(III) potassium sulphate
- Chromium(III) potassium sulphate-12-water
- Chromium(III) potassium sulphate dodecahydrate

. . . and various others!

You will also find variations on its formula. For example:

- $CrK(SO_4)_2,12H_2O$
- $Cr_2(SO_4)_3,K_2SO_4,24H_2O$
- $K_2SO_4,Cr_2(SO_4)_3,24H_2O$

The first of these formulae is just the other ones divided by two and rearranged a bit. Personally, I prefer the second one because it is easier to understand what is going on.

Chrome alum is known as a *double salt*. If you mix solutions of potassium sulphate and chromium(III) sulphate so that their molar concentrations are the same, the solution behaves just like you would expect of such a mixture. It gives the reactions of chromium(III) ions, of potassium ions, and of sulphate ions.

However, if you crystallise it, instead of getting mixed

crystals of potassium sulphate and chromium(III) sulphate, the solution crystallises as single deep purple crystals. These are "chrome alum".

Chrome alum crystals can be made by reducing acidified potassium dichromate(VI) solution using ethanol, and then crystallising the resulting solution.

Assuming you use an excess of ethanol, the main organic product will be ethanal - and we've already seen this equation above:

$$Cr_2O_7^{2-} + BH^+ + 3CH_3CH_2OH \longrightarrow 2Cr^{3+} + 7H_2O$$

$$+3CH_3CHO$$

This ionic equation obviously doesn't contain the spectator ions, potassium and sulphate. Feeding those back in gives the full equation:

$$K_2Cr_2O_7 + 4H_2SO_4 + 3CH_3CH_2OH \longrightarrow Cr_2(SO_4)$$

$$+K_2SO_4 + 7H_2O + 3CH_3CHO$$

If you look at the top line on the right-hand side of the equation, you will see that the chromium(III) sulphate and potassium sulphate are produced in exactly the right proportions to make the double salt.

What you do, then, is this:

You start with a solution of potassium dichromate(VI) to which has been added some concentrated sulphuric acid. The solution is then cooled by standing it in ice.

An excess of ethanol is added slowly with stirring so that the temperature doesn't rise too much.

When all the ethanol has been added, the solution is left over-night, preferably in a refrigerator, to crystallise. The crystals can be separated from the remaining solution, washed with a little pure water and then dried with filter paper.

Using Potassium Dichromate(VI) as an Oxidising agent in Titrations

Potassium dichromate(VI) is often used to estimate the concentration of iron(II) ions in solution. It serves as an alternative to using potassium manganate(VII) solution.

In Practice

There are advantages and disadvantages in using potassium dichromate(VI).

Advantages

- Potassium dichromate(VI) can be used as a primary standard. That means that it can be made up to give a stable solution of accurately known concentration. That isn't true of potassium manganate(VII).
- Potassium dichromate(VI) can be used in the presence of chloride ions (as long as the chloride ions aren't present in very high concentration).

Potassium manganate(VII) oxidises chloride ions to chlorine; potassium dichromate(VI) isn't quite a strong enough oxidising agent to do this. That means that you don't get unwanted side reactions with the potassium dichromate(VI) soution.

Disadvantage

- The main disadvantage lies in the colour change. Potassium manganate(VII) titrations are self-indicating. As you run the potassium manganate(VII) solution into the reaction, the solution becomes colourless. As soon as you add as much as one drop too much, the solution becomes pink - and you know you have reached the end point.

Unfortunately potassium dichromate(VI) solution turns green as you run it into the reaction, and there is no way you could possibly detect the colour change when you have one drop of excess orange solution in a strongly coloured green solution.

With potassium dichromate(VI) solution you have to use a separate indicator, known as a *redox indicator*. These change colour in the presence of an oxidising agent.

There are several such indicators - such as diphenylamine sulphonate. This gives a violet-blue colour in the presence of excess potassium dichromate(VI) solution. However, the colour is made difficult by the strong green also present. The end point of a potassium dichromate(VI) titration isn't as easy to see as the end point of a potassium manganate(VII) one.

The Calculation

The half-equation for the dichromate(VI) ion is:

$$Cr_2O_7^{2-} + 14H^+ + 6e^- \longrightarrow 2Cr^{3+} + 7H_2O$$

. . . and for the iron(II) ions is:

$$Fe^{2+} \longrightarrow Fe^{3+} + e^-$$

Combining these gives:

$$Cr_2O_7^{2-} + 14H^+ + 6Fe^{2+} \longrightarrow 2Cr^{3+} + 7H_2O + 6Fe^{3+}$$

You can see that the reacting proportions are 1 mole of dichromate(VI) ions to 6 moles of iron(II) ions.

Once you have established that, the titration calculation is going to be just like any other one.

Testing for Chromate(VI) ions in Solution

Typically, you would be looking at solutions containing sodium, potassium or ammonium chromate(VI). Most chromates are at best only slightly soluble; many we would count as insoluble.

The bright yellow colour of a solution suggests that it would be worth testing for chromate(VI) ions.

Testing by Adding an Acid

If you add some dilute sulphuric acid to a solution containing chromate(VI) ions, the colour changes to the familiar orange of dichromate(VI) ions.

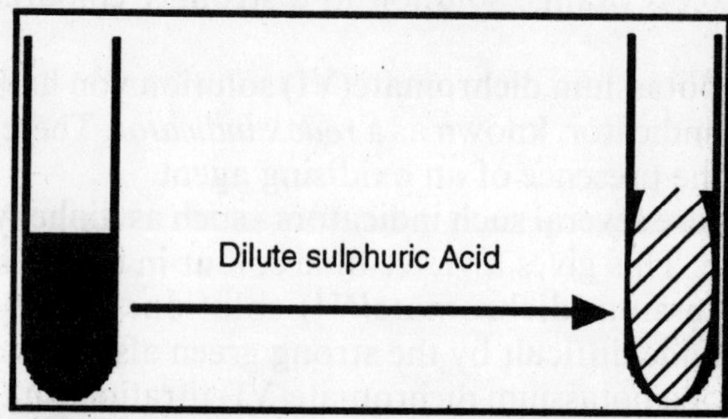

You can't rely on this as a test for chromate(VI) ions,

however. It might be that you have a solution containing an acid-base indicator which happens to have the same colour change!

Testing by adding barium chloride (or nitrate) solution

Chromate(VI) ions will give a yellow precipitate of barium chromate(VI).

$$Ba^{2+}{}_{(aq)} + CrO_4{}^{-2}{}_{(aq)} \longrightarrow BaCrO_{4(s)}$$

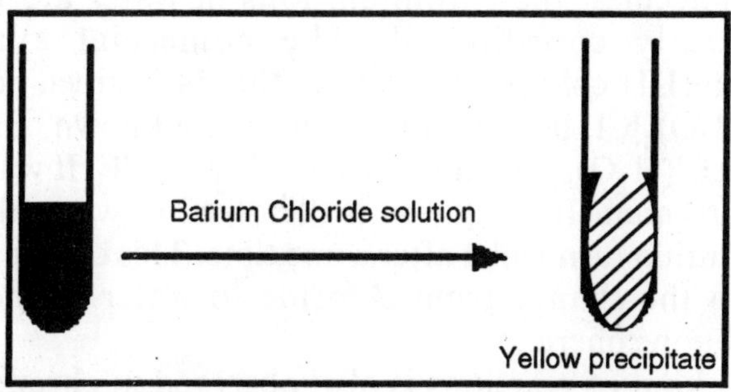

Testing by Adding lead(II) Nitrate Solution

Chromate(VI) ions will give a bright yellow precipitate of lead(II) chromate(VI). This is the original "chrome yellow" paint pigment.

$$Pb^{2+}{}_{(aq)} + CrO_4{}^{2-}{}_{(aq)} \longrightarrow PbCrO_{4(s)}$$

Lead (II) Nitrate Solution

Very bright yellow precipitate
"chrome yellow"

CHEMISTRY AND COMPOUNDS

The oxidation state 3+ is the most stable one and therefore

a large number of chromium(III) compounds is known. Chromium(III) can be obtained by dissolving chromium in acids like hydrochloric acid or sulfuric acid. The aluminium(III) (ion radius 0.50 E) and chromium(III) (ion radius 0.63 E) can replace each other in some compounds, for example chrome alum and alum. Another example is aluminium oxide (corundum, Al_2O_3) where by replacement, the red colored ruby is formed.

Chromium tends to form complexes, for example with water molecules; the chromium ions in water are usually octahedrally coordinated. The comercial available chromium(III) chloride hydrate is the dark green complex $[CrCl_2(H_2O)_4]Cl$, but two other forms are known, *viz.*, pale green $[CrCl(H_2O)_5]Cl_2$ and violet $[Cr(H_2O)_6]Cl_3$. If water free green chromium(III) chloride is dissolved in water then the green solution turn violet after some time. This colour change is due to the change from chloride to water in the inner coordination sphere.

This kind of reactions is also observed in chrome alum solutions and some other water soluble chromium(III) salts, and the reverse reaction can be induced by heating the solution. The chromium(III) hydroxide $Cr(OH)_3$ shows amphoteric reactions and dissolves in acid water by forming $[Cr(H_2O)_6]^{3+}$ and in basic water by forming $[Cr(OH)_6]^{3-}$. By heating the chromium(III) hydroxide it is transformed into the green chromium(III) oxide (Cr_2O_3), which is the stable oxide (melting point of 2275□°C) with the crystal structure identical to that of corundum.

The second stable oxidation state is 6+, for example the chromate, which is produced in large scale by oxidative roasting of chromite ore with calcium or sodium carbonate. Chromate and dichromate are in an equilibrium, which is influenced in that case by the law of mass action and therefore by the pH of the solution.

$$2\,CrO_4^{2-} + 2\,H_3O^+ \rightarrow Cr_2O_7^{2-} + 2\,H_2O$$

The change in equilibrium is also visible by a change from yellow (chromate) to orange (dichromate) if an acid is added to a neutral solution of potassium chromate. At lower pH, further condensation to more complex oxyanions of chromium is possible. The chromate and dichromate are strong oxidizing reagents at low pH

$Cr_2O_7^{2-} + 14\ H_3O^+ + 6\ e^- \rightarrow 2\ Cr^{3+} + 21\ H_2O\ \varepsilon_0 = 1.33V$

but only moderate ones at high pH

$CrO_4^{2-} + 4\ H_2O + 3\ e^- \rightarrow Cr(OH)_{3+} + 5\ OH^-\ \varepsilon_0 = -0.13V.$

Chromium(VI) compounds in solution can be detected by adding acidic hydrogen peroxide solution. A dark blue unstable chromium(VI) peroxide (CrO_5) is formed which can be stabilized as an ether adduct $CrO_5 \cdot OR_2$. Chromic acid has the hypothetical structure H_2CrO_4. Neither chromic nor dichromic acid is found in nature, but their anions are found in a variety of compounds, the chromates and dichromates.

Chromium(VI) oxide CrO_3, the acid anhydride of chromic acid, is sold industrially as "chromic acid".The dark red chromium(VI) oxide can be produced by mixing sulfuric acid with dichromate, and is an extremely strong oxidizing agent.

The oxidation state 5+ is only realized in few compounds. The only binary compound is the highly volatile chromium(V) fluoride (CrF_5).

This red solid with a melting point of 30°C and a boiling point of 117°C can be synthesized by reacting fluorine with chromium at 400°C and 200 bar pressure.

The peroxochromate(V) is another example of the oxidation state 5+. The potassium peroxochromate ($K_3[Cr(O_2)_4]$) is made by reacting potassium chromate with hydrogen peroxide at low temperatures. This red brown compound is stable at room temperature but decomposes spontaneously between 150–170°C.

The chromium(IV) compounds (4+) are slightly more stable than the chromium(V) compounds, and the halogen compounds CrF_4, $CrCl_4$ and $CrBr_4$ can be produced by the reaction of the trihalogens with additional elementary halogens at elevated temperatures. Most of the compounds are susceptible to disproportionation reactions and therefore are not stable in water. An example for a Chromium(II) compounds (2+) is the water stable chromium(II) chloride which can be produced by reduction of chromium(III) chloride with zinc. The resulting light blue solutions are only stable at neutral pH when the solution is very pure.

PRODUCTION

Approximately 4.4 million metric tons of marketable chromite ore were produced in 2000, and converted into ~3.3 million tons of ferro-chrome with an approximate market value of 2.5 billion United States dollars.[20] The largest producers of chromium ore have been South Africa (44%) India (18%), Kazakhstan (16%) Zimbabwe (5%), Finland (4%) Iran (4%) and Brazil (2%) with several other countries producing the rest of less than 10% of the world production.

The two main products of chromium ore refining are ferrochromium and metallic chromium.

For those products the ore smelter process differs considerably.

For the production of ferrochromium, the chromite ore ($FeCr_2O_4$) is reduced in large scale in electric arc furnace or in smaller smelters with either aluminium or silicon in an aluminothermic reaction.

For the production of pure chromium, the iron has to be separated from the chromium in a two step roasting and leaching process. The chromite ore is heated with a mixture of calcium carbonate and sodium carbonate in the presence of air. The chromium is oxidized to the hexavalent form, while the iron forms the stable Fe_2O_3. The subsequent leaching at higher elevated temperatures dissolves the chromates and leaves the insoluble iron oxide. The chromate is converted by sulfuric acid into the dichromate.

$$4FeCr_2O_4 + 8\ Na_2CO_3 + 7\ O_2 \rightarrow 8\ Na_2CrO_4 + 2\ Fe_2O_3 + 8\ CO_2$$

$$Na_2CrO_4 + H_2SO_4 \rightarrow Na_2Cr_2O_4 + Na_2SO_4 + H_2O$$

The dichromate is converted to the chromium(III) oxide by reduction with carbon and then reduced in an aluminothermic reaction to chromium.

$$Na_2Cr_2O_4 . H_2O + 2\ C \rightarrow Cr_2O_3 + Na_2CO_3 + 2\ H_2O + CO$$

$$Cr_2O_3 + 2\ Al \rightarrow Al_2O_3 + Cr$$

APPLICATIONS

Metallurgy

The strengthening effect on steel by forming stable carbide grains at the grain boundaries and the strong increase in corrosion resistance made chromium an important alloying material for steel. The high speed tool steels contain between 3 and 5% chromium. An important stainless steel is *18/10 stainless*, made from iron with 10% nickel and 18% chromium, is widely used for cookware and cutlery.

For these applications, ferrochromium is added to the molten iron. Also nickel based alloys increase in strength due to the formation of stable carbide grains at the grain boundaries. For example, Inconel 718 contains 18.6% chromium. Because of the excellent heat stability of these nickel superalloys, they are used in jet engines and gas turbines in large quantities.

The relative high hardness and corrosion resistance of unalloyed chromium makes it a good surface coating. A thin layer of chromium is deposited on pretreated metallic surfaces by electroplating techniques. There are two deposition methods: Thin, below 1 μm thickness, layers are deposited by chrome plating, and are used for decorative surfaces.

If wear-resistant surfaces are needed then thicker chromium layers of up to mm thickness are deposited. Both methods normally use acidic chromate or dichromate solutions. To prevent the energy consuming change in oxidation state, the use of Chromium(III) sulfate is under development, but for most applications, the established process is used.

In the chromate conversion coating process, the strong oxidative properties of chromates are used to deposit a protective oxide layer on metals like aluminium, zinc and cadmium. This passivation and the self healing properties by the chromate stored in the chromate conversion coating, which is capable to migrate to local defects, are the benefits of this coating method. Because of environmental and health regulations on chromates, alternative coating method are under development. Anodizing of aluminium is another electrochemical process, which does not lead to the deposition of chromium, but uses chromic acid as electrolyte in the solution. During anodization, an oxide layer is formed on the aluminium.

The use of chromic acid, instead of the normally used sulfuric acid, leads to a slight difference of these oxide layers. The high toxicity of Cr(VI) compounds, used in the established chromium electroplating process, and the strengthening of safety and environmental regulations demand a search for substitutes for chromium or at least a change to less toxic chromium(III) compounds.

Dye and Pigment

The mineral crocoite (lead chromate $PbCrO_4$) was used as a yellow pigment shortly after its discovery. After a synthesis method became available starting from the more abundant chromite, Chrome yellow was, together with cadmium yellow, one of the most used yellow pigments. The pigment is light stable and has a strong colour.

The signaling effect of yellow was used for school buses in the United States and for Postal Service (for example Deutsche Post) in Europe.

The use of chrome yellow declined due to environmental and safety concerns and was substituted by organic pigments or other lead-free alternatives. Other pigments based on chromium are, for example, the bright red pigment Chrome red, which is a basic lead chromate ($PbCrO_4.Pb(OH)_2$).Chrome green is a mixture of Prussian blue and chrome yellow, while the Chrome oxide green is Chromium(III) oxide.

Glass is colored green by the addition of chromium(III) oxide. This is similar to emerald, which is also colored by chromium.A red colour is achieved by doping chromium(III) into the crystals of corundum, which are then called ruby. Therefore, chromium is used in producing synthetic rubies.

The toxicity of chromium(VI) salts is used in the preservation of wood. For example, chromated copper arsenate (CCA) is used in timber treatment to prevent wood from decay fungi, wood attacking insects, including termites, and marine borers.The formulations contain chromium based on the oxide CrO_3 between 35.3% and 65.5%. In the United States, 65,300 metric tons of CCA solution have been used in 1996.

Tanning

Chromium(III) salts, especially chrome alum and chromium(III) sulfate, are used in the tanning of leather. The chromium(III) stabilizes the leather by cross linking the collagen fibers within the leather. Chromium tanned leather can contains between 4 and 5% of chromium, which is tightly bound to the proteins. Refractory material

The high heat resistivity and high melting point makes chromite and chromium(III) oxide a material for high temperature refractory applications, like blast furnaces, cement kilns, molds for the firing of bricks and as foundry sands for the casting of metals.

In these applications, the refractory materials are made from mixtures of chromite and magnesite. The use is declining because of the environmental regulations due to the possibility of the formation of chromium(VI).

Other use

Several chromium compounds are used as catalyst. For example the Phillips catalysts for the production polyethylene are mixtures of chromium and silicon dioxide or mixtures of chromium and titanium and aluminium oxide. Chromium(IV) oxide (CrO_2) is a magnetic compound.

Its ideal shape anisotropy, which imparted high coercivity and remanent magnetization, made it a compound superior to the γ-Fe_2O_3.

Chromium(IV) oxide is used to manufacture magnetic tape used in high performance audio tape and standard audio cassette. Chromates can prevent corrosion of steel under wet conditions, and therefore chromates are added to the drilling muds.

The long known influence of chromium uptake on diabetes conditions suggested the positive influence of dietary supplement containing chromium(III) also on healthy persons.

For this reason, dietary supplement or slimming aid usually contain chromium(III) chloride, chromium(III) picolinate, chromium(III) polynicotinate or amino acid chelate, such as chromium(III) D-phenylalanine.

The benefit of those supplements is still under investigation and is questioned by some studies.

Chapter 12

Cobalt

Cobalt is a hard, lustrous, grey metal, a chemical element with symbol Co and atomic number 27. Although cobalt-based colors and pigments have been used since ancient times, and miners have long used the name kobold ore for some minerals, the free metalic cobalt was not prepared and discovered until 1735 by Georg Brandt.

Cobalt is found in various metallic-lustred ores for example cobaltite (CoAsS), but it is produced as a by-product of copper and nickel mining. The copper belt in the Democratic Republic of the Congo and Zambia yields most of the worldwide mined cobalt.

Cobalt is used in the preparation of magnetic, wear-resistant, and high-strength alloys. Cobalt blue (cobalt(II) aluminate, $CoAl_2O_4$) gives a distinctive deep blue colour to glass, ceramics, inks, paints, and varnishes. Cobalt-60 is a commercially important radioisotope, used as a tracer and in the production of gamma rays for industrial use.

Cobalt is an essential trace-element for all multicellular organisms as the active centre of coenzymes called cobalamins. These include vitamin B-12 which is essential for mammals. Cobalt is also an active nutrient for bacteria, algae, and fungi, and may be a necessary nutrient for all life.

CHARACTERISTICS

Cobalt is a ferromagnetic metal. Pure cobalt is not found in nature, but compounds of cobalt are common. Small amounts of it are found in most rocks, soil, plants, and animals. It is the element of atomic number 27. The Curie temperature is 1388 K, and the magnetic moment is 1.6–1.7 Bohr magnetons per atom. In nature, it is frequently associated with nickel, and both are characteristic minor components of meteoric iron.

Mammals require small amounts of cobalt which is the basis of vitamin B_{12}. Cobalt-60, an artificially produced radioactive isotope of cobalt, is an important radioactive tracer and cancer-treatment agent. Cobalt has a relative permeability two thirds that of iron. Metallic cobalt occurs as two crystallographic structures: hcp and fcc. The ideal transition temperature between hcp and fcc structures is 722□K, but in practice, the energy difference is so small that random intergrowth of the two is common. Cobalt has a hardness of 5.5 on the Mohs scale of mineral hardness.

Isotopes

Cobalt is the only stable cobalt isotope. 22 radioisotopes have been characterized with the most stable being ^{60}Co with a half-life of 5.2714 years, ^{57}Co with a half-life of 271.79 days, ^{56}Co with a half-life of 77.27□days, and ^{58}Co with a half-life of 70.86 days. All of the remaining radioactive isotopes have half-lives that are less than 18□hours, and the majority of these are less than 1□second. This element also has 4 meta states, all of which have half-lives less than 15 minutes.

The isotopes of cobalt range in atomic weight from 50 u (^{50}Co) to 73 u (^{73}Co). The primary decay mode for isotopes with atomic mass unit values less than that of the most abundant stable isotope, ^{59}Co, is electron capture and the primary mode of decay for those of greater than 59 atomic mass units is beta decay. The primary decay products before ^{59}Co are element 26 (iron) isotopes and the primary products after are element 28 (nickel) isotopes.

Cobalt Radioisotopes in Medicine

Cobalt-60 (Co-60 or ^{60}Co) is a radioactive metal that is used in radiotherapy. It produces two gamma rays with energies of 1.17 MeV and 1.33MeV. The ^{60}Co source is about 2□cm in diameter and as a result produces a geometric penumbra, making the edge of the radiation field fuzzy. The metal has the unfortunate habit of producing a fine dust, causing problems with radiation protection. Cobalt-60 has a radioactive half-life of 5.27 years. This decrease in activity requires periodic replacement of the sources used in radiotherapy and is one reason why cobalt machines have been largely replaced by linear accelerators (linacs) in modern radiation therapy.

Cobalt-57 (Co-57 or ^{57}Co) is a radioactive metal that is used in medical tests. It is used as a radiolabel for vitamin B_{12} uptake, and is useful for the Schilling test

Industrial Uses for Radioactive Isotopes

Cobalt-60 (Co-60 or ^{60}Co) is useful as a gamma ray source because it can be produced in predictable quantity and high activity by simply exposing natural cobalt to neutrons in a reactor for a period.

Its uses include sterilization of medical supplies and medical waste, radiation treatment of foods for sterilization (cold pasteurization), industrial radiography (e.g., weld integrity radiographs), density measurements (e.g., concrete density measurements), and tank fill height switches. Cobalt-57 is used as a source in Mφssbauer spectroscopy and is one of several possible sources in XRF devices (Lead Paint Spectrum Analyzers).

REACTIONS OF COBALT(II) IONS IN SOLUTION

The simplest ion that cobalt forms in solution is the pink hexaaquacobalt(II) ion - $[Co(H_2O)_6]^{2+}$.

The reaction of hexaaquacobalt(II) ions with hydroxide ions

Hydroxide ions (from, say, sodium hydroxide solution) remove hydrogen ions from the water ligands attached to the cobalt ion. Once a hydrogen ion has been removed from two of the water molecules, you are left with a complex with no charge - a neutral complex. This is insoluble in water and a precipitate is formed.

$$\left[Co(H_2O)_6\right]^{2+} + 2OH^- \longrightarrow \left[Co(H_2O)_4(OH)_2\right] + 2H_2O$$

In the test-tube, the colour changes are:

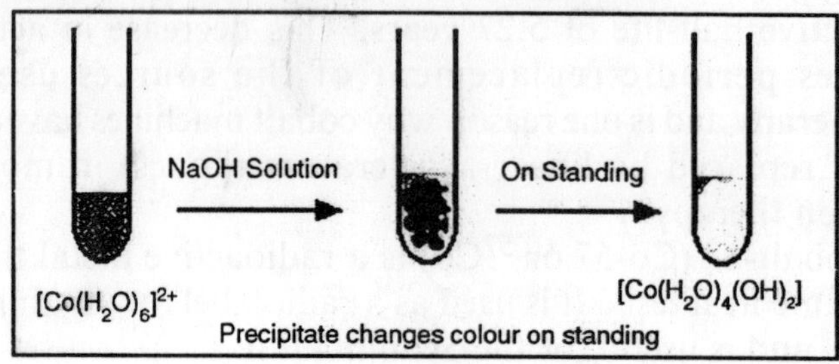

Precipitate changes colour on standing

Reactions of Hexaaquacobalt(II) Ions with Aammonia Solution

The ammonia acts as both a base and a ligand. With a small amount of ammonia, hydrogen ions are pulled off the hexaaqua ion exactly as in the hydroxide ion case to give the same neutral complex.

$$\left[Co(H_2O)_6\right]^{2+} + 2NH_3 \longrightarrow \left[Co(H_2O)_4(OH)_2\right] + 2NH_4^+$$

That precipitate dissolves if you add an excess of ammonia. The ammonia replaces water as a ligand to give hexaamminecobalt(II) ions.

$$\left[Co(H_2O)_6\right]^{2+} + 6NH_3 \longrightarrow \left[Co(NH_3)_6\right]^{2+} + 6H_2O$$

The colour changes are:

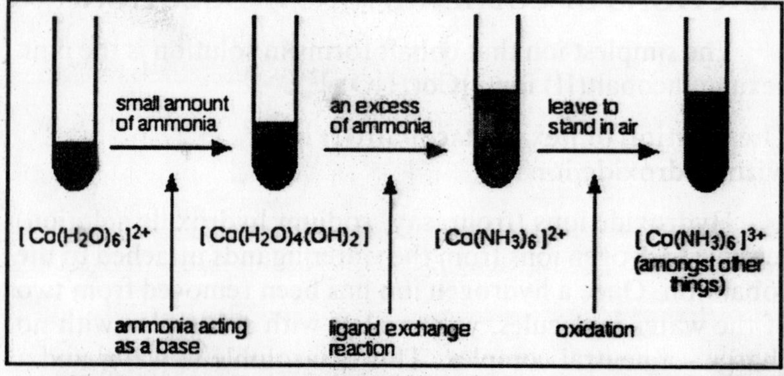

The hexaamminecobalt(II) complex is very easily oxidised

to the corresponding cobalt(III) complex. In the test-tube this is seen as a rapid darkening to a deep red-brown solution.

In fact the hexaamminecobalt(III) ion is *yellow!* What you see is a mixture of this ion and various other cobalt(III) ions involving ligand exchange reactions with both water molecules and negative ions present in the solution.

Another Oxidation of Hexaaquacobalt(II) ions

Rather than relying on oxidation by the air, you can add an oxidising agent such as hydrogen peroxide. You can do this after the addition of ammonia as in the last case, or you can do it following addition of sodium hydroxide solution.

$$2Co(NH_3)_6^{2+} + H_2O_2 \longrightarrow 2Co(NH_3)_6^{3+} + 2OH^-$$

With Ammonia Solution and Hydrogen Peroxide

The reaction with ammonia solution followed by hydrogen peroxide produces the same dark reddish-brown solution as before - only faster. The equation for the oxidation of the ammine complex is:

With Sodium Hydroxide Solution and Hydrogen Peroxide

You get the variably coloured precipitate of the cobalt(II) hydroxide complex when you add the sodium hydroxide solution.

Addition of hydrogen peroxide produces lots of bubbles of oxygen and a dark chocolate brown precipitate.

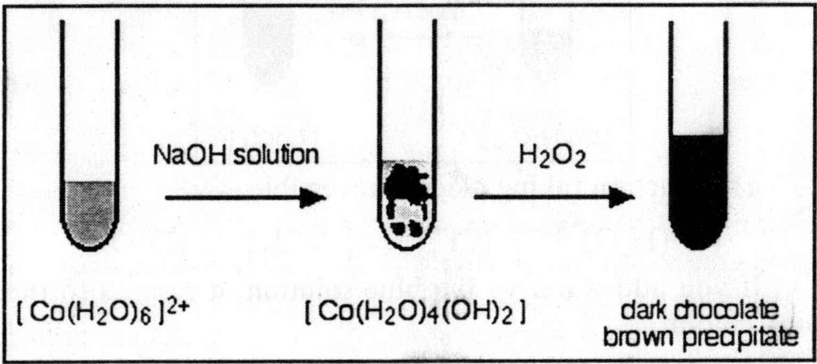

The final precipitate contains cobalt in the +3 oxidation state. I haven't given a formula for this, because I don't know exactly what it is! I can't give an equation for the same reason.

The Reaction of Hexaaquacobalt(II) ions with Carbonate ions

You simply get a precipitate of what you can think of as cobalt(II) carbonate.

$$Co^{2+}(aq) + CO_3^{2-}(aq) \longrightarrow CoCO_{3(s)}$$

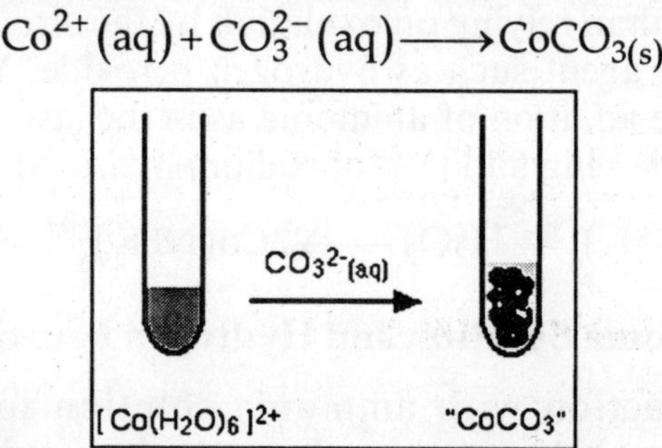

A Ligand Exchange Reaction Involving Chloride ions

If you add concentrated hydrochloric acid to a solution containing hexaaquacobalt(II) ions, the solution turns from its original pink colour to a rich blue. The six water molecules are replaced by four chloride ions.

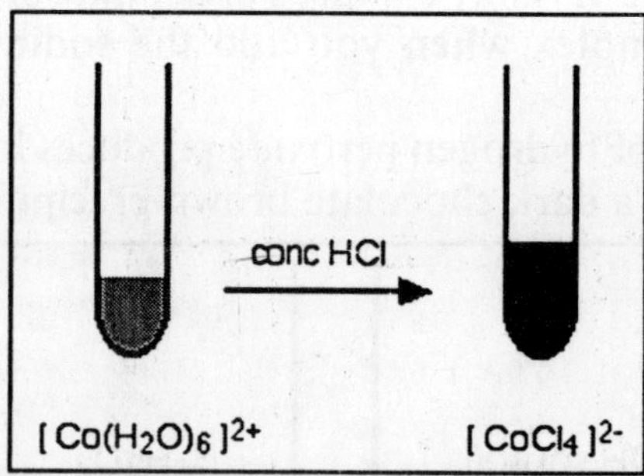

The reaction taking place is reversible.

$$[Co(H_2O)_6]^{2+} + 4Cl^- \rightleftharpoons [CoCl_4]^{2-} + 6H_2O$$

If you add water to the blue solution, it returns to the pink colour.

APPLICATIONS

Alloys

The cobalt based superalloys consume most of the produced cobalt. The temperature stability of these alloys make them suitable for turbine blades within gas turbines and jet aircraft engines. The nickel-based single crystal alloys surpass the cobalt based in temperature stability, but the cobalt based are still in use. These alloys are also corrosion and wear-resistant.

Special cobalt chromium molybdenum alloys are used for prosthetic parts such as hip and knee replacements. Some high speed steels also use cobalt to increase heat and wear-resistance.

The special alloys of aluminium, nickel and cobalt, known as Alnico, and of samarium and cobalt (samarium-cobalt magnet) are used in permanent magnets, which can be used for recording media, cemented carbides (also called hard metals), and diamond tools.

Batteries

Lithium cobalt oxide ($LiCoO_2$) is widely used in Lithium ion battery electrodes. Nickel-cadmium (NiCd) and nickel metal hydride (NiMH) batteries also contain significant amounts of cobalt.

Catalyst

Several cobalt compounds are used in chemical reactions as catalysts. Cobalt acetate is used for the production of terephthalic acid as well as dimethyl terephthalic acid, which are key compounds in the production of Polyethylene terephthalate. The steam reforming and hydrodesulfuration for the production of petroleum, which uses mixed cobalt molybdenum aluminium oxides as a catalyst, is another important application. Cobalt and its compounds, especially cobalt carboxylates (known as cobalt soaps), are good oxidation catalysts. They are used in paints, varnishes, and inks as drying agents through the oxidation of certain compounds. The same carboxylates are used to improve the adhesion of the steel to rubber in steel-belted radial tires.

Pigments and coloring

Before the 19th century, the predominant use of cobalt was the pigmentation of glass. The colors cobalt blue and cobalt green originated from this use.

Cobalt has been used to colour glass since the Bronze Age. The excavation of the Uluburun shipwreck yielded an ingot of blue glass which was cast during the 14th century BC. Blue glass items from Egypt are colored with copper, iron, or cobalt. The oldest cobalt-colored glass was from the time of the Eighteenth dynasty (1550–1292 BC). The location where the cobalt compounds were obtained is unknown.

Other uses

- Electroplating due to its appearance, hardness, and resistance to oxidation
- Ground coats for porcelain enamels
- Purification of histidine-tagged fusion proteins in biotechnology applications

Chapter 13

Copper

Copper is a chemical element with the symbol Cu (Latin: *cuprum*) and atomic number 29. It is a ductile metal with very high thermal and electrical conductivity. Pure copper is rather soft and malleable and a freshly-exposed surface has a pinkish or peachy colour. Gold, caesium and copper are the only metallic elements with a natural colour other than gray or white. It is used as a thermal conductor, an electrical conductor, a building material, and a constituent of various metal alloys.

Copper is an essential trace nutrient to all higher plant and animal life. In animals, including humans, it is found widely in tissues, with concentration in liver, muscle, and bone. It functions as a co-factor in various enzymes and in copper-based pigments. Some molluscs have blue-green blood resulting from a copper compound which they use to transport oxygen, instead of heme. In sufficient amounts, copper salts can be poisonous.

Copper has played a significant part in the history of mankind, which has used the easily accessible uncompounded metal and its alloys for thousands of years. Evidence has been preserved from several early civilizations of the use of copper. In the Roman era, copper was principally mined on Cyprus, hence the origin of the name of the metal as Cyprium, "metal of Cyprus", later shortened to Cuprum.

Some countries, such as Chile and the United States, still have sizable reserves of the metal which are extracted through large open pit mines. However, like tin, there may be insufficient reserves to sustain current rates of consumption.

High demand relative to supply caused a price spike in the 2000s.

Copper has a significant presence in decorative art. It can also be used as an anti-germ surface that can add to the anti-bacterial and antimicrobial features of buildings such as hospitals.

CHARACTERISTICS

Copper has a reddish, orangish, or brownish colour because a thin layer of tarnish (including oxides) gradually forms on its surface when gases (especially oxygen) in the air react with it. But pure copper, when fresh, is actually a pinkish or peachy metal. Copper, caesium and gold are the only three elemental metals with a natural colour other than gray or silver.[15] The usual gray colour of metals depends on their "electron sea" that is capable of absorbing and re-emitting photons over a wide range of frequencies. Copper has its characteristic colour because of its unique band structure.

By Madelung's rule the 4s subshell should be filled before electrons are placed in the 3d subshell but copper is an exception to the rule with only one electron in the 4s subshell instead of two. The energy of a photon of blue or violet light is sufficient for a *d* band electron to absorb it and transition to the half-full *s* band. Thus the light reflected by copper is missing some blue/violet components and appears red. This phenomenon is shared with gold which has a corresponding 5s/4d structure. In its liquefied state, a pure copper surface without ambient light appears somewhat greenish, a characteristic shared with gold. When liquid copper is in bright ambient light, it retains some of its pinkish luster.

When copper is burnt in oxygen it gives off a black oxid

Mechanical Properties

Copper is easily worked, being both ductile and malleable. The ease with which it can be drawn into wire makes it useful for electrical work in addition to its excellent electrical properties. Copper can be machined, although it is usually necessary to use an alloy for intricate parts, such as threaded components, to get really good machinability characteristics. Good thermal conduction make it useful for heatsinks and in heat exchangers. Copper has good corrosion resistance, but not as good as gold. It has excellent brazing and soldering properties and can also be welded, although best results are obtained with gas metal arc welding.

Copper is normally supplied, as with nearly all metals for industrial and commercial use, in a fine grained polycrystalline form. Polycrystalline metals have greater strength than monocrystalline forms, and the difference is greater for smaller grain (crystal) sizes. The reason is due to the inability of stress dislocations in the crystal structure to cross the grain boundaries. The plastic deformation of polycrystal is similar to mild steel.

Electrical Properties

At 59.6×10^6 S/m copper has the second highest electrical conductivity of any element, just after silver. This high value is due to virtually all the valence electrons (one per atom) taking part in conduction.

The resulting free electrons in the copper amount to a huge charge density of 13.6×10^9 C/m^3. This high charge density is responsible for the rather slow drift velocity of currents in copper cable (drift velocity may be calculated as the ratio of current density to charge density).

For instance, at a current density of 5×10^6 A/m^2 (typically, the maximum current density present in household wiring and grid distribution) the drift velocity is just a little over S! mm/s.

Corrosion

Copper is a metal that does not react with water (H_2O), but the oxygen of the air will react slowly at room temperature to form a layer of brown-black copper oxide on copper metal.

It is important to note that in contrast to the oxidation of iron by wet air that the layer formed by the reaction of air with copper has a protective effect against further corrosion. On old copper roofs a green layer of copper carbonate, called verdigris, can often be seen. A notable example of this is on the Statue of Liberty

Isotopes

Copper has 29 distinct isotopes ranging in atomic mass from 52 to 80. Two of these, ^{63}Cu and ^{65}Cu, are stable and occur naturally, with ^{63}Cu comprising approximately 69% of naturally occurring copper.

The other 27 isotopes are radioactive and do not occur naturally. The most stable of these is ^{67}Cu with a half-life of 61.83 hours. The least stable is ^{54}Cu with a half-life of approximately 75 ns.

Unstable copper isotopes with atomic masses below 63 tend to undergo β^+ decay, while isotopes with atomic masses above 65 tend to undergo $\beta^{"}$ decay. ^{64}Cu decays by both β^+ and $\beta^{"}$.

^{68}Cu, ^{69}Cu, ^{71}Cu, ^{72}Cu, and ^{76}Cu each have one metastable isomer. ^{70}Cu has two isomers, making a total of 7 distinct isomers. The most stable of these is ^{68m}Cu with a half-life of 3.75 minutes. The least stable is ^{69m}Cu with a half-life of 360 ns.

REACTIONS OF COPPER(II) IONS IN SOLUTION

The simplest ion that copper forms in solution is the typical blue hexaaquacopper(II) ion - $[Cu(H_2O)_6]^{2+}$.

The Reaction of Hexaaquacopper(II) ions with Hydroxide ions

Hydroxide ions (from, say, sodium hydroxide solution) remove hydrogen ions from the water ligands attached to the copper ion.

Once a hydrogen ion has been removed from two of the water molecules, you are left with a complex with no charge - a neutral complex. This is insoluble in water and a precipitate is formed.

$$[Cu(H_2O)_6]^{2+} + 2OH^- \longrightarrow \left[Cu(H_2O)_4(OH)_2\right] + 2H_2O$$

In the test-tube, the colour change is:

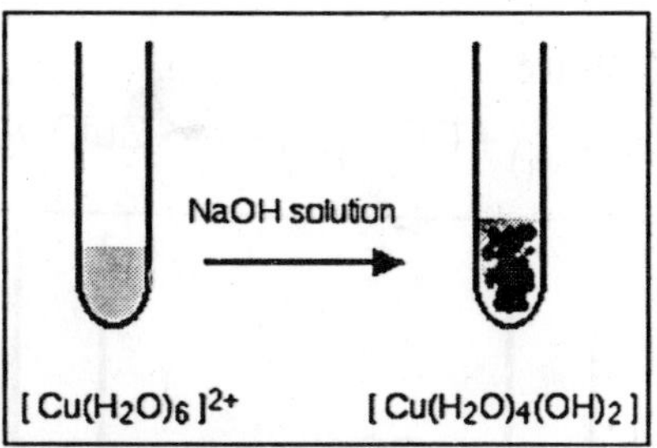

Reactions of Hexaaquacopper(II) ions with Ammonia Solution

The ammonia acts as both a base and a ligand. With a small amount of ammonia, hydrogen ions are pulled off the hexaaqua ion exactly as in the hydroxide ion case to give the same neutral complex.

$$[Cu(H_2O)_6]^{2+} + 2NH_3 \longrightarrow \left[Cu(H_2O)_4(OH)_2\right] + 2NH_4^+$$

That precipitate dissolves if you add an excess of ammonia. The ammonia replaces water as a ligand to give tetraamminediaquacopper(II) ions. Notice that only 4 of the 6 water molecules are replaced.

$$[Cu(H_2O)_6]^{2+} + 4NH_3 \longrightarrow \left[Cu(NH_3)_4(H_2O)_2\right]^{2+} + 4H_2O$$

The colour changes are:

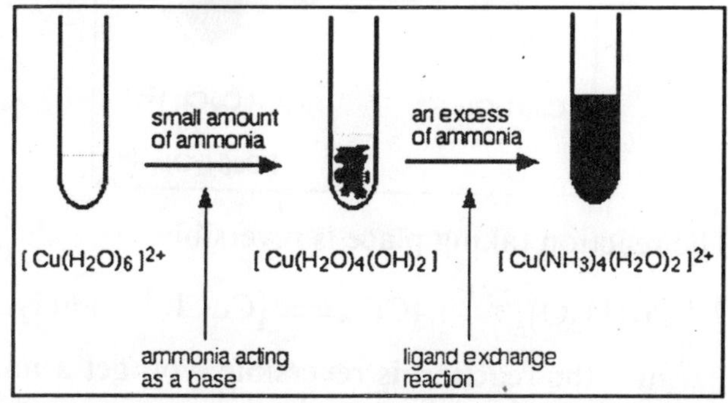

The Reaction of Hexaaquacopper(II) ions with Carbonate ions

You simply get a precipitate of what you can think of as

copper(II) carbonate.

$$Cu^{2+}{}_{(aq)} + CO_3^{2-}{}_{(aq)} \longrightarrow CuCO_{3(s)}$$

CO_3^{2-}(aq)

$[Cu(H_2O)_6]^{2+}$ "$CuCO_3$"

A Ligand Exchange Reaction Involving Chloride ions

If you add concentrated hydrochloric acid to a solution containing hexaaquacopper(II) ions, the six water molecules are replaced by four chloride ions.

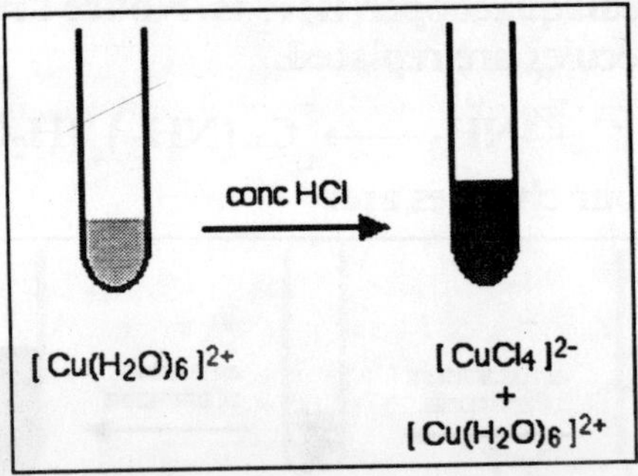

The reaction taking place is reversible.

$$\left[Cu(H_2O)_6\right]^{2+} + 4Cl^- \rightleftharpoons [CuCl_4]^{2-} + 6H_2O$$

Because the reaction is reversible, you get a mixture of colours due to both of the complex ions.

You may find the colour of the tetrachlorocuprate(II) ion variously described as olive-green or yellow.

If you add water to the green solution, it returns to the blue colour.

The Reaction of Hexaaquacopper(II) ions with iodide ions

The Simple Reaction

Copper(II) ions oxidise iodide ions to iodine, and in the process are themselves reduced to copper(I) iodide.

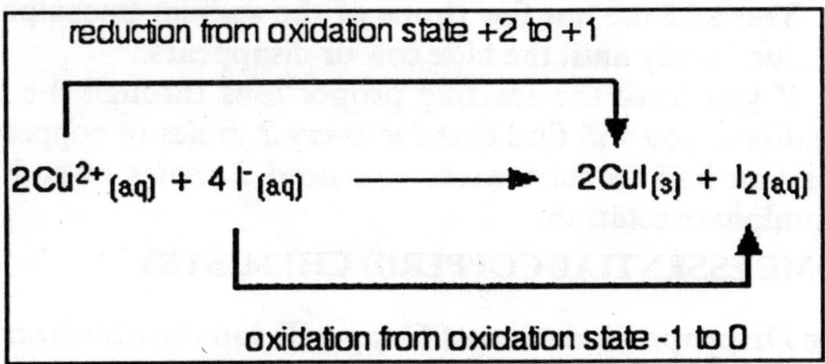

The initial mucky brown mixture separates into an off-white precipitate of copper(I) iodide under an iodine solution.

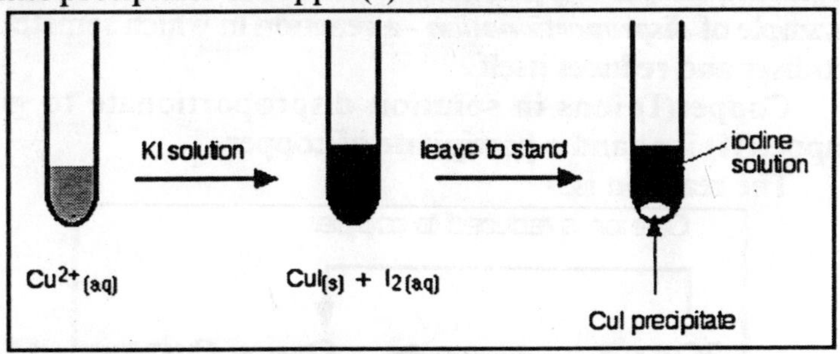

Using this Reaction to find the Concentration of Copper(II) ions in Solution

If you pipette a known volume of a solution containing copper(II) ions into a flask, and then add an excess of potassium iodide solution, you get the reaction we have just described.

$$2Cu^{2+}{}_{(aq)} + 4I^{-}{}_{(aq)} \longrightarrow 2CuI_{(s)} + I_{2(aq)}$$

You can find the amount of iodine liberated by titration with sodium thiosulphate solution.

$$2S_2O_3^{2-}{}_{(aq)} + I_{2(aq)} \longrightarrow S_4O_6^{2-}{}_{(aq)} + 2I^{-}{}_{(aq)}$$

As the sodium thiosulphate solution is run in from a burette, the colour of the iodine fades. When it is almost all gone, you add some starch solution. This reacts reversibly with iodine to give a deep blue starch-iodine complex which is much easier to see.

You add the last few drops of the sodium thiosulphate solution slowly until the blue colour disappears.

If you trace the reacting proportions through the two equations, you will find that for every 2 moles of copper(II) ions you had to start with, you need 2 moles of sodium thiosulphate solution.

SOME ESSENTIAL COPPER(I) CHEMISTRY

The Disproportionation of Copper(I) ions in Solution

Copper(I) chemistry is limited by a reaction which occurs involving simple copper(I) ions in solution. This is a good example of *disproportionation* - a reaction in which something oxidises and reduces itself.

Copper(I) ions in solution disproportionate to give copper(II) ions and a precipitate of copper.

The reaction is:

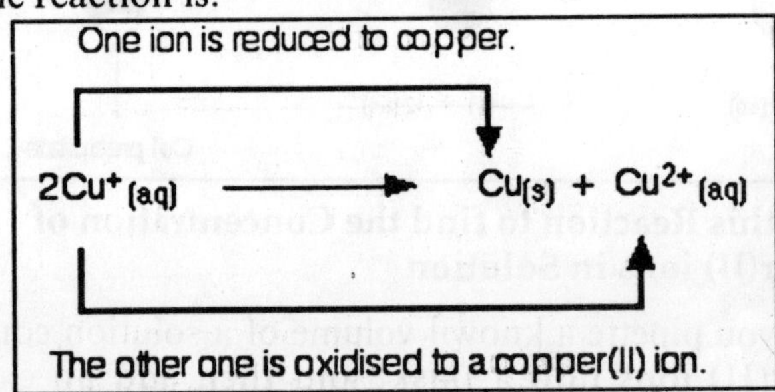

Any attempt to produce a simple copper(I) compound in solution results in this happening.

For example, if you react copper(I) oxide with hot dilute sulphuric acid, you might expect to get a solution of copper(I) sulphate and water produced.

In fact you get a brown precipitate of copper and a blue solution of copper(II) sulphate because of the disproportionation reaction.

$$Cu_2O + H_2SO_4 \longrightarrow Cu + CuSO_4 + H_2O$$

Stabalising the Copper(I) Oxidation State

Insoluble copper(I) compounds[ssh]

We've already seen that copper(I) iodide is produced as an off-white precipitate if you add potassium iodide solution to a solution containing copper(II) ions. The copper(I) iodide is virtually insoluble in water, and so the disproportionation reaction doesn't happen.

Similarly copper(I) chloride can be produced as a white precipitate (reaction described below). Provided this is separated from the solution and dried as quickly as possible, it remains white. In contact with water, though, it slowly turns blue as copper(II) ions are formed.

The disproportionation reaction only occurs with simple copper(I) ions in solution.

Copper(I) Complexes[ssh]

Forming copper(I) complexes (other than the one with water as a ligand) also stabalises the copper(I) oxidation state.

For example, both $[Cu(NH_3)_2]^+$ and $[CuCl_2]^-$ are copper(I) complexes which don't disproportionate.

The chlorine-containing complex is formed if copper(I) oxide is dissolved in concentrated hydrochloric acid.

You can think of this happening in two stages. First, you get copper(I) chloride formed:

$$Cu_2O_{(s)} + 2HCl_{(aq)} \longrightarrow 2CuCl_{(s)} + H_2O_{(l)}$$

But in the presence of excess chloride ions from the HCl, this reacts to give a stable, soluble copper(I) complex.

You can get the white precipitate of copper(I) chloride (mentioned above) by adding water to this solution. This reverses the last reaction by stripping off the extra chloride ion.

APPLICATIONS

Copper is malleable and ductile and is a good conductor of both heat and electricity.

The purity of copper is expressed as 4 N for 99.99 % pure or 7 N for 99.99999% pure.

The numeral gives the number of nines after the decimal point when expressed as a decimal (e.g. 4 N means 0.9999, or 99.99%).

Copper is often too soft for its applications, so it is incorporated in numerous alloys. For example, brass is a copper-zinc alloy, and bronze is a copper-tin alloy.

Index

D

E

F

G

H

I